Developments in Geotectonics 3

ISLAND ARCS

JAPAN AND ITS ENVIRONS

Further Titles in this Series

Developments in Geotectonics 3

ISLAND ARCS
JAPAN AND ITS ENVIRONS

BY

A. SUGIMURA

Research Associate
Geological Institute, The University of Tokyo, Tokyo, Japan

S. UYEDA

Professor
Earthquake Research Institute, The University of Tokyo, Tokyo, Japan

ELSEVIER SCIENTIFIC PUBLISHING COMPANY
Amsterdam - London - New York 1973

ELSEVIER SCIENTIFIC PUBLISHING COMPANY
335 JAN VAN GALENSTRAAT
P.O. BOX 1270, AMSTERDAM, THE NETHERLANDS

AMERICAN ELSEVIER PUBLISHING COMPANY, INC.
52 VANDERBILT AVENUE
NEW YORK, NEW YORK 10017

Library of Congress Card Number: 77-180008

ISBN 0-444-40970-X

WITH 134 ILLUSTRATIONS AND 11 TABLES

PRINTED IN THE NETHERLANDS

PREFACE

In Chapter 1, the geophysical and geological features of island arcs are reviewed, taking the arcs of the Japanese Islands as an example. The features consist of the topography, gravity, crust and upper mantle structure, seismicity, direction of principal stresses of earthquakes, crustal deformation and fault systems, geological structure, distributions of Cenozoic volcanoes and hot springs, petrology of volcanic rocks, terrestrial heat flow, magnetic field, and anomaly of electrical conductivity. Many of these features are arranged in zones parallel to the arcs. It is shown that these zonal arrangements are common to the island arcs in general. Some paleomagnetic data with possible relevance to island arc tectonics are also reviewed.

In Chapter 2, the Cenozoic history of the Japanese Islands in relation to the formation of the island arcs is described. Since the major relief of the arcs is considered to have resulted from Quaternary tectonic movements, the Quaternary tectonics of Japan are reviewed in detail. However, the youngest orogenic cycle started at the beginning of the Miocene along the belt of the present island arcs. Consequently, the Late Cenozoic history is important here. The orogenies of the latest Cretaceous to the Paleogene in southwest Japan and Hokkaido are also outlined as examples of the past island arc features.

In Chapter 3, island arc tectonics based on the assumption of an upper-mantle convection current are developed to account for the characteristic features dealt with in Chapter 1 and 2. The mantle current, with an overlying lithosphere, is assumed to descend in the trench area along the deep seismic zone. General low heat flow associated with the descending current in the oceanward side of island arcs appears to be enhanced by the occurrence of endothermic regional metamorphism. The descending mantle material drags down the surface layer to form the trench and the gravity anomaly, and the associated mechanical stresses cause shallow earthquakes. Oceanic lithosphere that is composed of oceanic crust, possibly including a part of surface sediments, and its low attenuating substratum, is dragged down into the mantle. Intermediate deep shocks may be due to Raleigh–Paterson type brittle rupture while deeper earthquakes may be due to Griggs–Orowan type plastic instability. The latter type earthquakes occurring at depths greater than ca. 130 km would produce magma by shear melting of either the buried crust or the mantle. The magma, thus produced, would ascend through the process of zone partial melting as suggested by Kushiro. This process will form the present surface distribution of volcanoes and the volcanic front. The process of zone partial melting is effective in heating the surrounding materials during the passage of magma, and this will be the origin of the general high heat flow in the inner zone of island arcs and the seas behind them. Most of the ascending magma cannot reach the surface under these seas because of the greater depths of generation. Numerical experiments suggest that the amount of molten materials in this model is large enough to form the upper mantle and floor of the marginal seas, such as the Sea of Japan. These seas might have originated by this type of spreading. Zonal distribution of petrological characteristics of volcanic rocks can be accounted for by the proposed process, and is also supported by laboratory experiments. Anomaly of electrical conductivity in the upper mantle may be explained by the combined effect of the ocean water and the subterranean temperature distribution estimated from heat flow data. The distribution of the horizontal component of principal stress associated with deep earthquakes is explained by the regional stress field. However, to account for the directions of stress axes in the vertical section some special consideration is needed. Here, an anisotropic nature of the mantle with respect to fault formation is hypothesized on the basis of Kumazawa's idea. But the alternative view that the stress distribution reflects the sinking of the lithospheric slab is also favoured. Anomalies of the geomagnetic field in the seas directly at ocean-side of island arcs appear to be related to the motion of the ocean floor in the remote past.

Tectogenetically, island arcs and mid-oceanic ridges are the representatives of the most active parts of the earth's surface: it is suggested that these two features represent the descending and ascending portions of the mantle currents, or the source and sink of the spreading earth's surface, which are the principal agents of the tectogenesis. Finally, an extended paired-belt concept is proposed for the island arc type or Pacific type orogeny and a model compatible with the ideas of the new global tectonics has been developed. In this model, reproduction of marginal sea plates is proposed as the fourth important process, in addition to the plate production at ridges, consumption at trenches and slippage along transform faults.

Present synthesis of island arcs started on March the sixth of 1964 when the authors happened to have a luncheon together at the campus cafeteria at the University of Tokyo and agreed, with excitement, that descending of mantle flow (subduction in today's language) could be substantiated strictly from the island arc studies. Sugimura had presented his model of descending movement along an inclined zone in his 1958 paper given the hint by the 1956 paper of Dr. T. Rikitake. Uyeda had worked out on the geothermal studies in Japan and environs since 1957 with Dr. K. Horai and had been getting a definite idea on the dynamic process under the area. The start of this synthesis in 1964 was a few years before the advent of plate tectonics. Ever since, however, the development of the new global tectonics has been so rapid and our pace so slow that the authors regret that the book may suffer from being outdated here and there. Nevertheless, it is the hope of the authors that their original excitement is somewhat preserved in this book.

We owe a great debt of gratitude to the assistance and discussion of many of our colleagues. We should thank Prof. A. Miyashiro, who stimulated our study of island arcs. We also thank all who allowed us to include their figures. Miss T. Tanaka, M. Mitani, T. Asamura, and N. Mizushima assisted the authors in drawing diagrams and typing manuscript. The research and writing of this book was done by Uyeda at the University of Tokyo and by Sugimura at the same university and at the State University of New York at Binghamton.

CONTENTS

CHAPTER 1

Geophysical and Geological Features of Island Arcs

During the last decade, information on the structure and activity of island arcs has greatly increased, in part through the general progress of the studies of the earth's crust and upper mantle, and in part through the intensive investigations made specifically on island arcs. In particular, the advent of the sea-floor spreading hypothesis and the new global tectonics appears to have brought about a revolutionary progress in the solid earth sciences, including island arc studies.

In this chapter an investigation of data on various aspects of the geophysics and geology of island arcs is made. It was intended to put as many kinds of information as possible into maps of the same projection. Throughout the present volume, data from the arcs in Japan and its environs are used as an example of the general island arc features. In this way, the present review follows those by Otuka (1938), Hess (1948) and Rikitake (1956) and is distinguished from those classical reviews such as by Umbgrove (1947), Jacobs et al. (1959) and Vening Meinesz (1964) in which the East and West Indies are dealt with more extensively.

ISLAND ARCS IN THE WORLD

The Pacific Ocean is surrounded, at least around its western and south-eastern margins, by belts of seismicity and volcanism: the circum-Pacific island arc—trench belts. It has long been noted that in the island arcs there is a distinct regularity in the arrangement of crust—mantle features, suggesting that all the island arc—trench systems have been brought about by a common mechanism. Many theories or hypotheses have been postulated on this mechanism. Among the older theories are those of Sollas (1903), Molengraaff (1914), Argand (1916), Hobbs (1925), Lake (1931) and Lawson (1932). Typical examples of the more modern theories are the earth's contraction hypotheses (Jeffreys, 1952; Wilson, 1959), down-buckling and convection current hypotheses (Vening Meinesz, 1930, 1964; Kuenen, 1936; Umbgrove, 1938, 1947; Griggs, 1939), serpentinization hypothesis (Hess, 1937), mantle fault hypotheses (Ewing and Heezen, 1955). In the last decade, the mantle convection hypothesis seems to have acquired favour among geo-scientists.

There are two major active systems over the globe: the mid-oceanic ridge and rift system and the orogenic mountain and island arc system (Fig.1). The circum-Pacific island arcs are the major members of the latter system and are more active at present than the Mediterranean—Himalayan arcuate mountain system. The "active" island arcs are

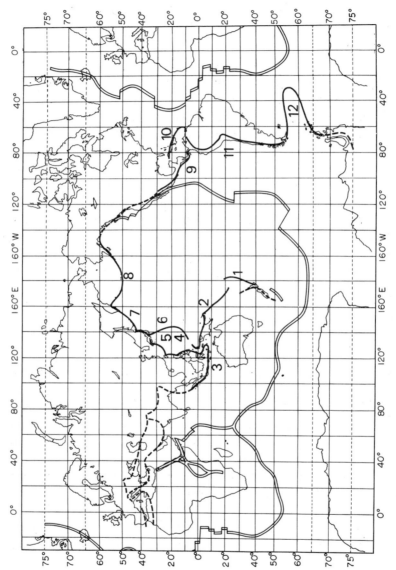

Fig.1. World rift system (double line), island arc system (thick line) and Alpine orogenic zone (broken line). Rift system is displaced at many places by fracture zones (thin line). For the numbers attached to the island arcs, see text.

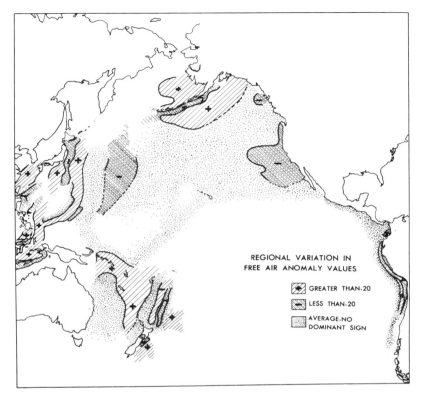

Fig.2. Free-air gravity anomaly of the Pacific Basin. (After Woollard and Strange, 1962.)

decidedly anomalous areas of the earth, having the following major characteristics (Fig. 2–7):

(1) Arcuate continuation of islands.

(2) Prominent volcanic activity at present (Fig.3).

(3) Deep trench on the oceanic side (Fig.7) and shallow tray-shaped seas on the continental side.

(4) Marked gravity anomaly belt that indicates large departures from isostasy (Fig.2).

(5) Active seismicity, including deep and intermediate earthquakes (Fig.4, 5).

(6) Earth movement in progress.

(7) Coincidence of arcs with recent orogenic belts.

In recent years, some further characteristic features such as the distribution of heat flow (Fig.6), the composition of volcanic rocks and so forth, which also show remarkable zonalities, have become known to us.

Not all of the island arcs have been investigated with respect to each of the above characteristics, but upon the three criteria: (a) recent volcanic activity; (b) oceanic trenches deeper than 6,000 m (Fisher and Hess, 1963); and (c) earthquake foci deeper

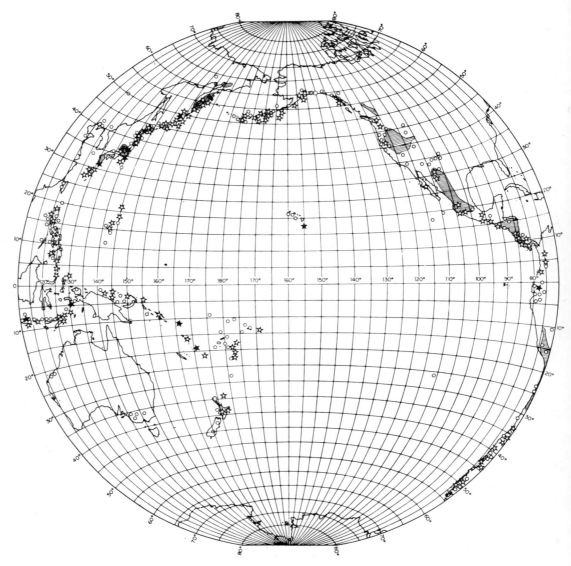

Fig.3. Distribution of volcanoes and Cenozoic volcanics. (Arranged from Vening Meinesz, 1964; and Lee and Uyeda, 1965.) ★ = active volcano; ☆ = dormant volcano; ○ = extinct volcano; ◉ = Cenozoic volcanics.

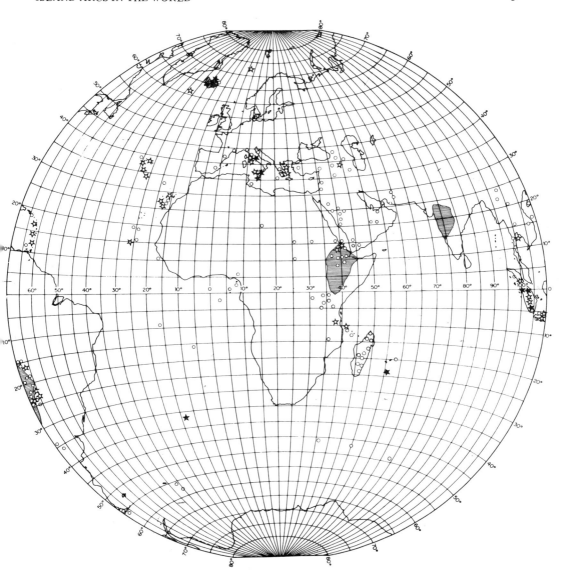

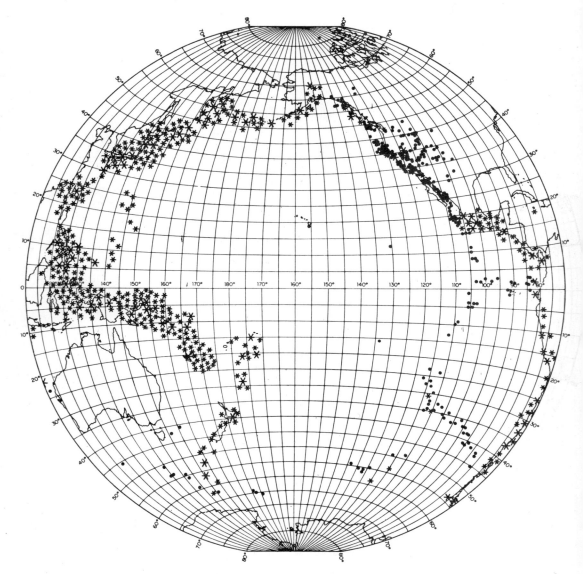

Fig.4. Distribution of shallow focus earthquake epicenters. (Arranged from Gutenberg and Richter, 1954.) For oceanic areas, except island arc areas, and western North America, 1966 epicenters due to U.S.C.G.S. are plotted (black circles).

★:Magnitude = 7–7.7 (1918–1952); ✗ :Magnitude > 7¾ (1904–1952).

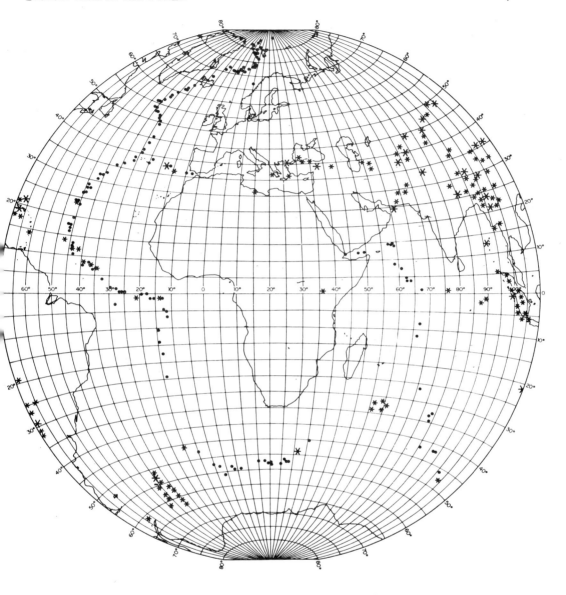

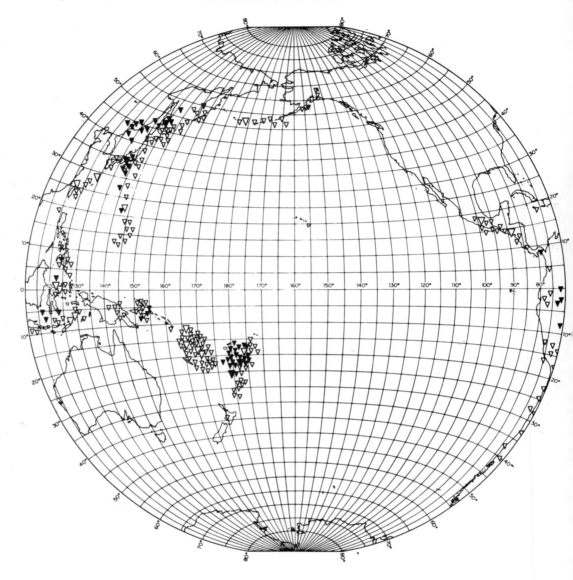

Fig.5. Distribution of deep focus earthquake epicenters (1904–1952). (Arranged from Gutenberg and Richter, 1954.)

▽ : magnitude = 7–7.7 }
▽ : magnitude > 7¾ } intermediate earthquakes (depth = 70–300 km)
▼ : magnitude = 7–7.7 }
▼ : magnitude > 7¾ } deep earthquakes (depth > 300 km)

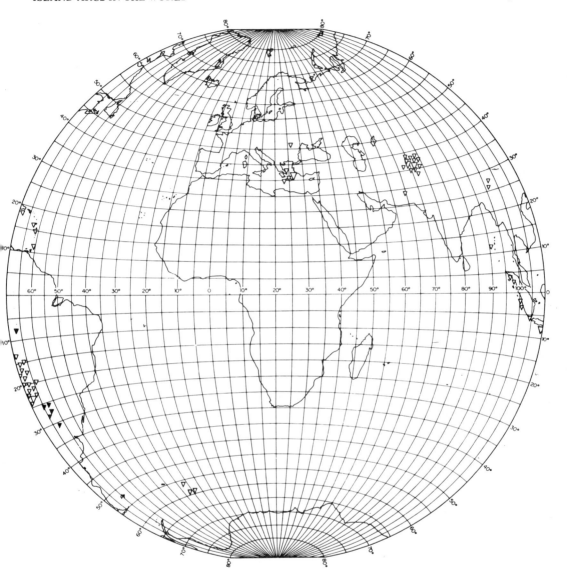

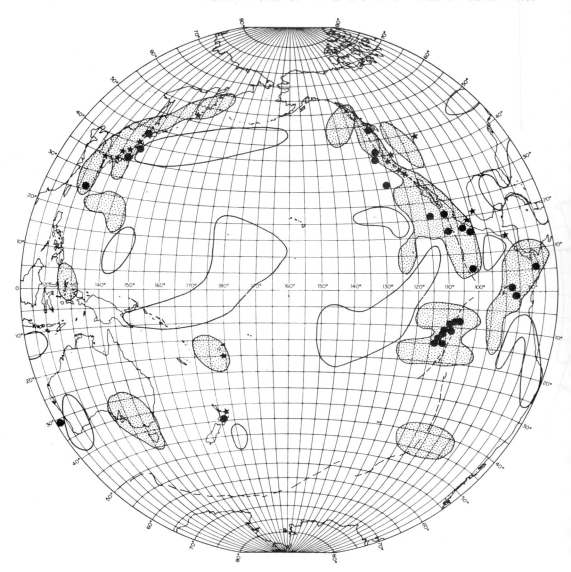

Fig.6. Anomaly of heat flow. (Mainly after Lee and Uyeda, 1965.) ◉ = regionally higher than $2.0 \cdot 10^{-6}$ cal/cm^2 sec; ○ = regionally lower than $1.0 \cdot 10^{-6}$ cal/cm^2 sec; ● = value higher than $5 \cdot 10^{-6}$ cal/cm^2 sec; − − − = rift system; ★ = geothermal area.

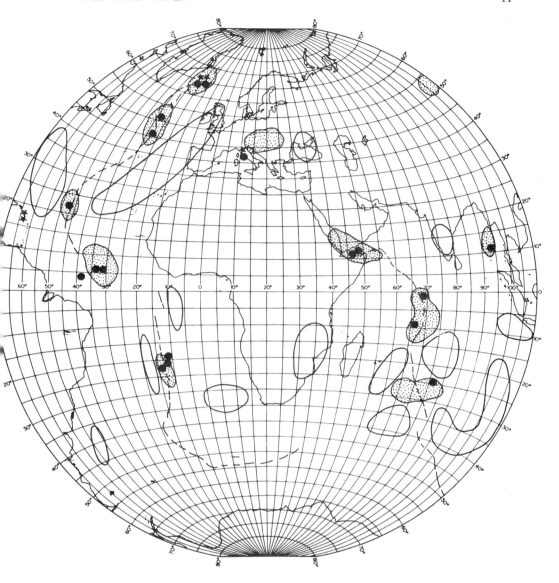

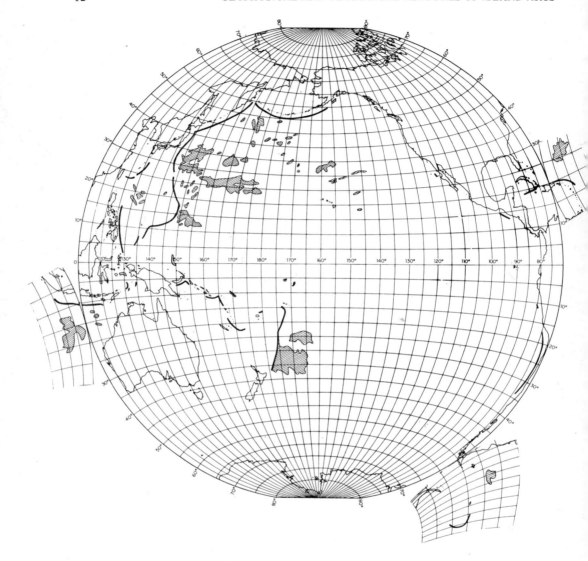

Fig.7. Distribution of the sea deeper than 6,000 m. Thick lines indicate trenches.

than 70 km (Gutenberg and Richter, 1954), the following may be identified as island arcs (see Fig.1): (*1*) New Zealand to Tonga; (*2*) Melanesia; (*3*) Indonesia; (*4*) Philippines; (*5*) Formosa (Taiwan) and west Japan; (*6*) Marianas and east Japan; (*7*) Kurile and Kamchatka; (*8*) Aleutian and Alaska; (*9*) Central America; (*10*) West Indies; (*11*) South America; and (*12*) western Antarctica (Sugimura, 1967b).

Among these arcs, Central and South America are not islands, but they are included in the list because they appear to have most of the other characteristic features. Each island arc has a length of the order of several thousands of kilometres with a narrow width (200–300 km including the oceanic trench).

The circum-Pacific belt does not perfectly surround the Pacific Ocean, as can be seen from Fig.1. As to the continuity of the belt all around the Pacific, Girdler (1964) cast doubts, on the basis of the non-occurrence of deep and intermediate earthquakes in western North America and the Antarctic coast. The ocean-floor spreading hypothesis and the new global tectonics (e.g., Vine, 1966; Isacks et al., 1968) seem to provide a logical explanation for the breaks of the circum-Pacific belt in these regions. However, the existence of a continued belt of "active" arcs in the western and southeastern rims of the Pacific is still highly impressive. A general index map of the northwestern Pacific area is given in Fig.8. ·

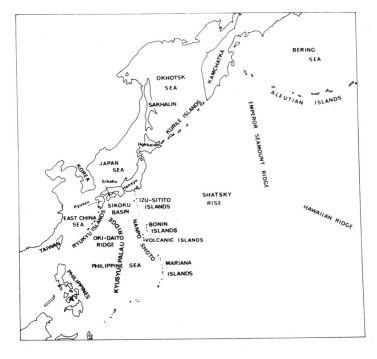

Fig.8. Index map of the northwestern Pacific area. Geographic names appearing in the text are indicated.

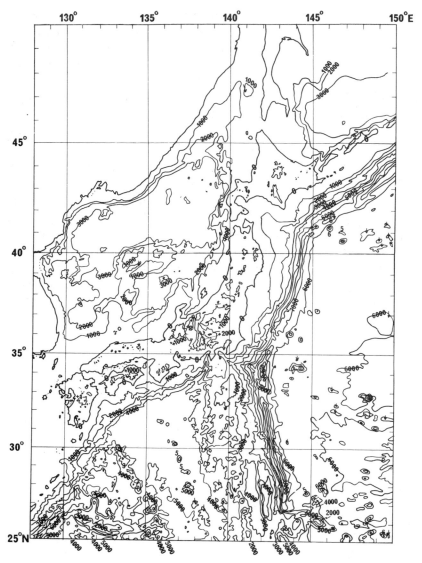

Fig.9. Topography of Japan and environs contoured in 1,000 m. (Arranged from Chart No. 6301–6304, Maritime Safety Board of Japan, 1966; and the Geographical Survey Institute of Japan, 1957a.)

TOPOGRAPHY

The physiographic features of the island arcs and the oceanic trenches bordering the Asiatic continent and the Pacific Ocean are spectacular. Fig.9 is a map showing the topography of Japan and its environs by contours at 1,000-m intervals. As is apparent in

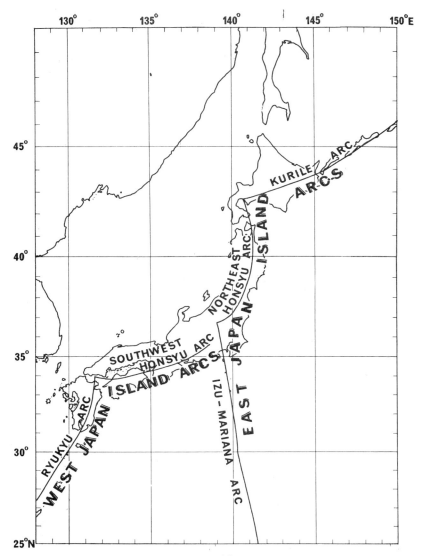

Fig.10. Grouping of the island arc systems of Japan.

this figure, the chain of trenches extending from Kurile to the Marianas is the most prominent feature. There is another trench system along the Ryukyu Islands. Thus, the authors consider that the more significant way of grouping the arcs should be as indicated in Fig.10: the arcs are grouped into two systems, i.e., the east Japan island arcs and the west Japan island arcs. In this grouping, Honsyu Island is divided into two: the Northeast Honsyu Arc and the Southwest Honsyu Arc. The Japan Trench runs parallel to the Northeast Honsyu Arc and curves toward the Izu—Ogasawara and Mariana trenches along the

TABLE I

Depths of the trenches around Japan (after Sato, 1969)

Trench	Depth (m)
Mariana	11,034
Izu—Ogasawara (Bonin)	9,810
Kurile—Kamchatka	9,783 or 9,550
Japan	8,412
Ryukyu (Nansei Syoto)	7,881

Izu—Mariana Arc. In the north, it meets the southwestern end of the Kurile Trench. These trenches form the principal foredeep of the east Japan island arcs. South of Kyusyu, the Ryukyu Trench runs along the Ryukyu Arc.

The oceanic trenches are without doubt the most important of the island arc features. Table I shows the maximum depths of the trenches around Japan. Fisher and Hess (1963) listed the maximum depths of the world's trenches. Most of these depths are shown in Table IV (see p.82). Table I differs from their list in the depths of the Kurile—Kamchatka Trench and the Ryukyu Trench. The maximum value of 7,507 m for the Ryukyu Trench is given by Fisher and Hess (1963) and differs by only a small amount from the value in Table I. The difference for the Kurile—Kamchatka Trench is large, as they show the depth of 10,542 m for this trench. The R.V. "Takuyo" made a detailed sounding in 1962 around the "deepest" point of 10,542 m found by the R.V. "Vitiaz". The interval of the courses for sounding was 5—10 km over an area of more than 5,000 km^2, and 2—6 km for the deepest part, which consists of a flat bottom of 9,500—9,550 m. The R.V. "Takuyo" could not detect a place deeper than 9,550 m. Nitani and Imayoshi (1963) and Iwabuchi (1968) reported the procedure and result, and suggested that the R.V. "Vitiaz" sounding was probably incorrect. Sato (1969) concluded that the greatest depth in the Kurile—Kamchatka Trench was 9,783 m at another place, but this was also open to question.

The Ryukyu Trench and Nankai Trough form the foredeep of the west Japan island arcs. The latter depression along the Southwest Honsyu Arc was named the Nankai Trough by Tayama (1950) and the southwest Japan Trench by Hoshino (1963). It is shallow (5,337 m) compared with other trenches. S. Murauchi and others (personal communication, 1967) found sediments more than 1,500 m thick in this trough south of the Muroto Cape. If the sediments were removed, the Nankai Trough would show a trench topography (T. Sato, personal communication, 1967). The Southwest Honsyu Arc and the Nankai Trough may be older than other nearby arc—trench systems. On the other hand, Hilde et al. (1969) have suggested a contrasting view, based on seismic profiling in the area, that the Nankai Trough is a juvenile trench.

The northern end of the west Japan island arcs strikes against the back of the east Japan island arcs in central Japan, where geological features are complicated (see p.50).

It may be that the west Japan arcs had extended further northeastward and were later cut by the younger east Japan arcs.

Among the above-mentioned arcs, the Ryukyu, Izu–Bonin–Mariana and a part of the Kurile arcs are of double arc topography. From the ocean-side towards the continent, features are characterized by a trench, a non-volcanic arc and a volcanic arc. Such an arrangement is known to be common to other double arcs, the Indonesian Islands being typical. The Mariana Arc is exceptional since the outer arc is volcanic and the inner non-volcanic (Hess, 1948). The arrangement of arcs in the Marianas is interpreted as a result of extensional opening of the trough between arcs (Karig, 1970). The Northeast Honsyu Arc would look like a typical double arc if the sea level were raised by a few hundred metres (Miyashiro, 1967). Kitakami and Abukuma Mountains (Fig.63, see p.92) would form the non-volcanic outer arc, and the belt of volcanoes (Fig.38, see p.56) the inner arc.

Continentward of the island arcs, there are marginal seas, such as the Sea of Okhotsk the Sea of Japan, the China Sea and the Philippine Sea. The southern half of the Sea of Japan has a complex topography: there are large topographical highs such as Yamato Bank, Kita–Yamato Bank and Korea Plateau. The northern half of the Sea of Japan has, on the other hand, a flat deep basin. The Sea of Okhotsk is shallow in the north and deep in the south. Waters shallower than 200 m prevail in the China Sea. But, again, there is a trough (the Okinawa Trough) behind the Ryukyu Arc. The Philippine Sea also has a complicated topography: there is a submarine ridge between Kyusyu and Palau. The topography of these marginal seas has been described by Mogi (1969) who suggested that these seas were originated by extensional opening.

Far out in the northwestern Pacific, there is a large topographical uplift called the Shatsky Rise or the Northwest Pacific Rise at about 158°E 33°N. Further east, a prominent chain of seamounts called the Emperor Seamount Ridge extends from Kamchatka to the Milwaukee Bank, the northwestern end of the Hawaiian Ridge system (Fig.8).

GRAVITY

Gravity measurements on land started in Japan in 1880, and by 1915 there were over 120 stations. Since the advent of the spring gravity meters, more thorough surveys have been carried out by Tsuboi et al. (1953, 1954a, b, 1955, 1956a, b, c) and the Geographical Survey Institute (1955, 1956, 1957b, 1964, 1966). In the ocean area, some 70 submarine pendulum measurements on Nanpo–Syoto of the Izu–Mariana Arc and in the sea east of Japan were made by Matuyama (1936). Recently Worzel (1965a), and Y. Tomoda (personal communication, 1965) compiled all the pendulum values. Hagiwara (1967) made terrain corrections on all the then existing land data to produce the Bouguer gravity anomaly map of the Japanese area as shown in Fig.11. In this figure, Hagiwara's corrected mean Bouguer anomalies are used on land, while for the sea area, Tomoda–Worzel's contour

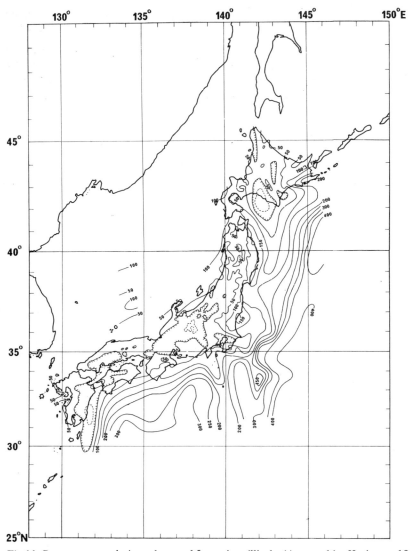

Fig.11. Bouguer anomaly in and around Japan in milligals. (Arranged by Hagiwara, 1967.)

map of terrain uncorrected Bouguer anomalies is used together. On the whole, the anomalies are small in magnitude in southwestern Japan, except for the area of the central mountains of Honsyu and eastern Kyusyu. The negative anomaly in the central mountains of Honsyu can be accounted for by an isostatic thickening of the crust. In northeastern Japan the distribution of the Bouguer anomaly is slightly more regular and zonal, and the anomaly increases seaward. A negative anomaly exists at the southern tip of Hokkaido. A less remarkable bulge of negative anomaly can be observed also in the southern Kanto area (including the Tokyo area). Together with the negative anomaly in the east of Kyusyu, the localities of these negative anomalies coincide with those of the intersections of two arcs. These negative anomalies found at the intersection of arcs may be of some tectonic significance (Matsuda and Uyeda, 1970). Hagiwara (1967) computed the short wavelength Bouguer anomaly in Japan as shown in Fig.12. In this map, anomalies with wavelength greater than about 200 km have been filtered out. Again in

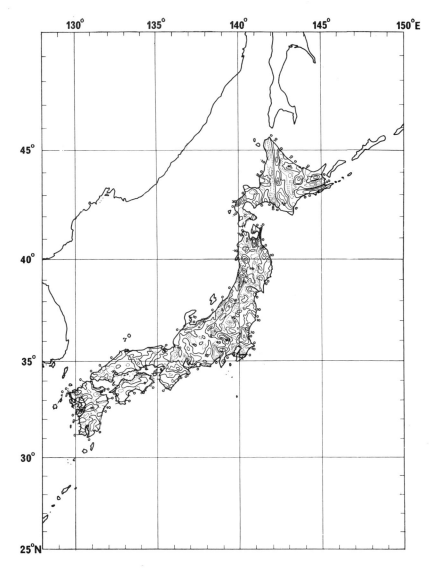

Fig.12. Short wavelength Bouguer anomaly in Japan. (After Hagiwara, 1967.) Contour interval is 10 milligals.

northeastern Japan the anomaly has a zonal pattern. A close correlation with geological structures can be observed in this map.

The Bouguer anomaly (Fig.11) increases oceanward to the value of 400 mgal in the Pacific proper, corresponding to the transition from the continental crust to the oceanic crust. It may be observed, however, that the area of the deep trench has a positive Bouguer anomaly greater than that of the Pacific Basin. This indicates that the general correlation between altitude and Bouguer anomaly expected from isostasy does not hold in that area. This situation can be seen clearly by the isostatic anomaly map of Fig.13 (D

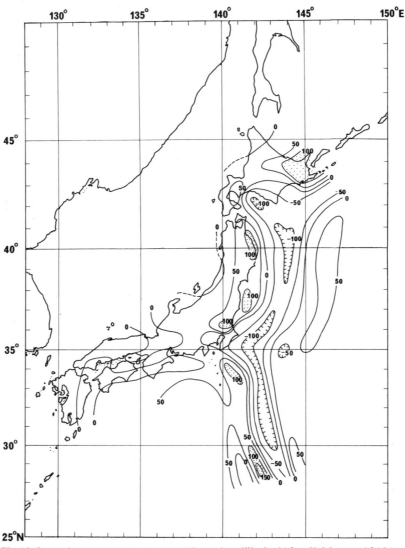

Fig.13. Isostatic anomaly in and around Japan in milligals. (After Heiskanen, 1945.)

= 40 km, R = 0 km, where D and R are the standard crust thickness and the regionality value, respectively, (Heiskanen, 1945; Heiskanen and Vening Meinesz, 1958). The trench area is out of the isostatic equilibrium by about 120 mgal. The positive anomaly zones on the Bonin Ridge, on the Pacific side of northeast Honsyu (Kitakami and Abukuma Mountains) and in eastern Hokkaido are the other remarkable features. Matsuda (1964) calls this belt of positive gravity anomaly the P-belt in contrast to the outer belt of negative anomaly called the N-belt. He considered that the combination of an N-belt and a P-belt forms the outer two belts of typical island arcs. The inner volcanic belt is called the I-belt. Usually the P and I belts form the double arcs, the outermost N-belt being taken as the trench. However, we might point out that the axis of the N-belt is almost always slightly closer to the islands than the topographical axis of the trench.

The great departure from the isostatic equilibrium in the Japan Trench areas corresponds to the classical observations made in the Indonesian Arc areas (see, e.g., Vening Meinesz, 1964) on which the hypothesis of downbuckling of the crust under the trench by a mantle convection current was based. The negative anomaly belt over the trench and its inner side has been illustrated on the Peru–Chile Trench (Ewing et al., 1957), the Tonga Trench (Talwani et al., 1961), the Middle America Trench (Worzel and Ewing, 1952), the Kurile–Kamtchatka Trench (Gainanov, 1955), the Mindanao Trench (Worzel and Shurbet, 1955), the Puerto Rico Trench (Talwani et al., 1959), etc. Whether the negative belt is really due to the downbuckling of the crust by compression or to the vertical drop of the crust by tensional forces (Worzel, 1965b) is not clear from these data alone. It seems, however, certain that some forces other than isostasy has brought about the existence of the ocean trenches and that these forces may be still working at the present time to maintain the anomaly. Maintenance of the negative anomaly by the strength of the crust may be conceivable (McKenzie, 1967), but this would not explain how such a non-equilibrium state was brought about at all. Tsuboi et al. (1961), Tomoda and Segawa (1967), Segawa (1968) and Tazima (1966a) have started extensive observations by surface-ship gravimeters that they themselves developed. Tomoda et al. (1968, 1970) have published a free-air anomaly map of the seas around Japan using surface ship data (Fig.14). Notable features such as the negative anomaly belt parallel to the trench and the negative anomalies at the intersections of arcs also are well observed in the figure. The Sea of Japan has essentially zero free-air anomaly just as has the Pacific Ocean basin. Considering the difference in the depth of these seas, J. Segawa (personal communication, 1970) suggests a thicker crust or lighter upper mantle under the Sea of Japan than under the Pacific.

For the studies of local or regional subterranean structures, observed gravity values, g, must be reduced properly (e.g., Heiskanen and Vening Meinesz, 1958). First, the effect of the height of the station, h, from the geoid (essentially the sea-level) must be corrected by $g_0 = g + 0.3086h$, where h is in metres and g in milligal. Then, the standard gravity on the geoid, γ_0, is subtracted, leaving the anomaly of gravity, called the free-air anomaly, $g_f = g_0 - \gamma_0$. The free-air anomaly has still to be reduced for the attraction of the mass

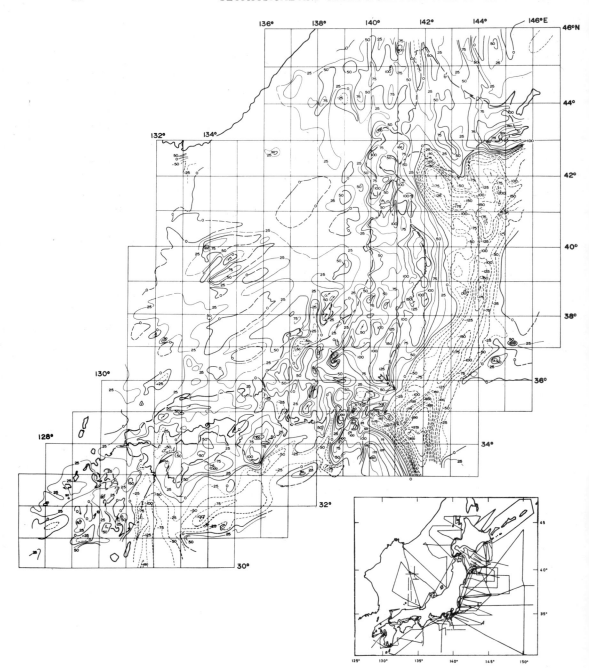

Fig.14. Free-air anomaly in and around Japan. (After Tomoda et al., 1970.) Figure at the right bottom shows the ship's tracks where gravity was measured.

between the sea-level and the station. The correction is given as $g_B = -2\pi k\rho_c h = -0.1118$ h mgal, where k is the gravitational constant and ρ_c (2.67) is the mean density of the mass. The net correction of the station height, then, becomes $+0.1968\ h$ mgal. $g_f + g_B$ is the Bouguer anomaly. The Bouguer anomaly must have its origin below the sea-level and therefore it should be a useful source of information about the underground mass distribution. Since the above approximation of the mass between the sea-level and the station by an infinite slab is crude, it is better to take note of the topography effect. This is called the terrain correction. Terrain correction is always positive.

In the oceanic area, gravity measurement is made essentially on the geoid, so that the observed gravity (g_o) subtracted by the standard gravity (γ_o) is the free-air anomaly and this should be equivalent to the Bouguer anomaly on land. But, when the water depth, d, is known, the deficiency of gravity due simply to the existence of water in place of crustal rocks can be corrected by adding $g_{B'} = 2\pi k(\rho_c - \rho_w)d = 0.0432d$ mgal, where ρ_w is the density of sea water (1.03). In the sea area, therefore, $g_f + g_{B'}$ is called the Bouguer anomaly. The Bouguer anomaly at sea is thus supposed to reflect the mass anomaly strictly below sea-bottom.

It is well known that the large scale topography is generally in the isostatically compensated state: that is, in the mountain area the Bouguer anomaly is negative and in the oceans it is positive. If we assign the depth of compensation, D, we can define the "bottom topography" of the Airy-type crust for the continent and ocean, respectively, by:

$$h_c \cdot \rho_c = H_c \cdot \Delta\rho \tag{1}$$

and:

$$d(\rho_c - 1.03) = H_o \cdot \Delta\rho \tag{2}$$

where $\Delta\rho$ is the density difference between the substratum and the surface stratum, and other quantities are defined in Fig.15. From the thus defined bottom topography of the surface stratum, the surface gravity due to the mass between the D_{min} level and D_{max} level can be computed. This will be the Bouguer anomaly in the case of a perfect isostatic

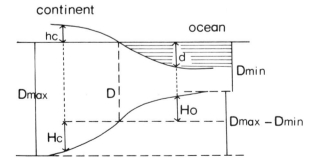

Fig.15. Explanation of isostatic compensation.

compensation. The difference between the observed Bouguer anomaly and this theoretical anomaly is called the isostatic anomaly. If, instead of the mass in columns, the vertical attraction is computed at each point at sea-level for the topography and its compensation, it will be properly compared with the observed terrain corrected Bouguer anomaly. Many models and calculations of the isostatic anomalies have been proposed. But, as can be seen in the above description, isostatic reduction requires idealized models with fixed values of D and $\Delta\rho$. These values, especially D, are indeterminate. If, however, $D = 0$, i.e., in the fictive case where the level of compensation is zero, the isostatic anomaly becomes equal to the free-air anomaly. Therefore, it may be said that practically all the important features of the elaborate isostatic anomaly can be inferred from the simple free-air anomaly.

CRUSTAL AND UPPER MANTLE STRUCTURE

Crustal structure of the Japanese area has been investigated by various methods. The Research Group for Explosion Seismology has been actively engaged on refraction studies. Their results for eight long profiles shown in Fig. 16 have been published in a series of papers in the *Bulletin of the Earthquake Research Institute* (Research Group for Explosion Seismology, 1951–1966). According to these studies and numerous models proposed, the crust under the Japanese Islands may be summarized as follows:

(*1*) a Vp = 5.5–5.8 km/sec layer less than 10 km thick usually occupies the uppermost crust.

(*2*) Underneath the first layer, a Vp = 6.0–6.5 km/sec layer extends with a thickness of 10–40 km. This second layer may be composed of two distinct layers: one with Vp = 6.0 km/sec, and the other with Vp = 6.5 km/sec. These are often called the granitic and basaltic layers. The details of the layered structure of the second layer are still unknown.

(*3*) The Vp of the uppermost mantle (*Pn*) under Japan is said to be anomalously low: i.e., Pn = 7.8–7.9 km/sec. It has been suggested recently, however, that Pn = 8.0 km/sec may be a better value (S. Asano, personal communication, 1968). The low value seems to come from Pn under east Japan and Pn under west Japan reaches to 8.4 km/sec (I. Muramatu, personal communication, 1967), but details are still to be clarified in the future.

Oceanic explosion seismology in Japan was initiated in 1962. Since then the work has been carried out actively by the cooperation of Japanese and American scientists (Murauchi et al., 1964, 1967, 1968; Ludwig et al., 1966; Den et al., 1969). Their major results are illustrated in Fig.17 and Fig.18. As can be seen in these figures, the deeper basin in the Sea of Japan has an oceanic crust, verifying the earlier Russian work (Andreyeva and Udintsev, 1958; Kovylin and Neprochnov, 1965; Kovylin, 1966). The area of Yamato Bank and Kita Yamato Bank has a thickened crust. The cross-section of the Japan Trench shows no sign of a thickened crust as postulated by the downbuckling hypothesis. But it does not show notable thinning (anti-root) of the crust either. At any

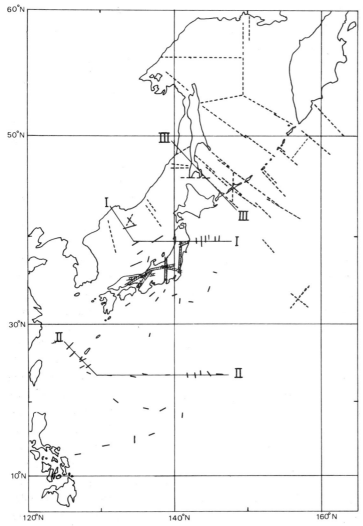

Fig.16. Distribution of seismic profiles. (Arranged from Murauchi and Yasui, 1968; and Mikumo, 1966.) Eight profiles over Japan are by the Research Group for Explosion Seismology (1951–1966).– are profiles by Murauchi et al. (1964, 1966, 1967) and by Ludwig et al. (1966); ––– are profiles taken by the U.S.S.R. Group (Kosminskaya et al., 1963). The lines I–I, II–II and III–III indicate the location of cross-sections in Fig.17–19.

rate, the existence of a greatly thickened crust as postulated from the tectogene theory is not supported in the Japan Trench area. In this respect, one recalls that Talwani et al. (1959) showed that Puerto Rico Trench is characterized by a thinning of the crust on both sides. Ludwig et al. (1966) maintain that the faults in the sediments of the slope of the Japan Trench are normal (see Fig.95, p.141). These observations appear to support

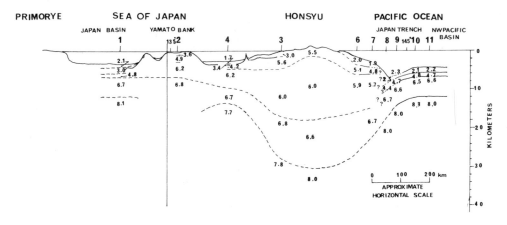

Fig.17. Crustal cross-section across the northeastern Japan along the line I–I in Fig.16. (After Murauchi and Yasui, 1968.)

the view that tensional forces are dominant in the crust of the trench area (e.g., Heezen, 1967). But the present authors are inclined to think that these tensional features are only superficial and that the regional force underneath is predominantly compressional (see Fig.28). It appears (Fig.18) that the ridges such as the Oki–Daito Ridge and Honsyu–Bonin Ridge have a thickened crust, with some Vp = 6.0 km/sec layer, and that the Philippine Sea has a rather complex structure. Complexities of the crust in the Philippine Sea may be related to those of its magnetic anomaly and heat flow (see pp.64 and 66). The Sikoku–Philippine Basin is not a typically normal ocean: it is a sea bordered by arcs on both sides. Toward the south, the crust in the Philippine Sea appears to become more normal–oceanic. The crustal structures of the Sea of Okhotsk and the Kurile–Kamchatka zone have been intensively studied by Russian scientists (Kosminskaya et al., 1963; Kosminskaya and Zverev, 1968; Gainanov et al., 1968; Tuyesov et al., 1968; Fig.16). The south Okhotian Basin, like the Japan Basin, has a sub-oceanic crust, whereas in the northern Okhotsk where the water depth is less than 2,000 m the crust is continental. A notable thickening of the crust was reported under continentward of the Kurile Trench, as shown in Fig.19. According to Den et al. (1969), Shatsky Rise in the northwest Pacific Basin shows a crustal layering appreciably different from the surrounding basin area. Under the rise, the crust is thickened to about 20 km by the occurrence of a layer with Vp = 7.3–7.8 km/sec between the third layer and the upper mantle.

Kanamori (1963) produced a map showing the most probable depth of the M-discontinuity in Japan. He used the map of the mean Bouguer anomaly (see p.17) from which the reduced Bouguer anomaly, ΔG, was computed by subtracting the effects due to the neighbouring crust. ΔG should be related to the depth of the M-discontinuity, D, by:

$$D = H_S - \frac{H_I (\rho_M - \rho_I)}{\rho_M - \rho_C} - \frac{\Delta G}{2\pi k^2 (\rho_M - \rho_C)} \qquad (3)$$

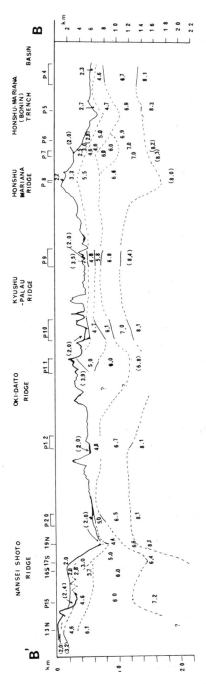

Fig.18. Crustal cross-section of the Philippine Sea along the line II–II in Fig.16. (After Murauchi and Yasui 1968.)

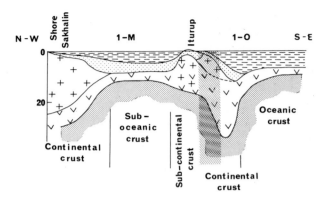

Fig.19. Crustal section across the Kurile Arc. (After Kosminskaya et al., 1963.) The location of the cross-section is indicated in Fig.16 by III–III.

where H_S is the thickness of the standard crust where ΔG is zero; H_I is the thickness of the intermediate layer directly below M-discontinuity ($Vp \cong 7.5$ km/sec): a layer supposedly characteristic to the uppermost mantle under Japan; and ρ_M, ρ_C, and ρ_I are the densities of the standard mantle, standard crust and the intermediate layer. Explosion and gravity studies indicated that the values of $H_S = 33$ km, $\rho_M = 3.27$ and $\rho_C = 2.84$, given by Worzel and Shurbet (1955), fit the case of Japan well. Assuming $H_I(\rho_M - \rho_I) = 1$ km g/cm^3, D was computed and contoured as shown in Fig.20 for each 1° square on the land. The depth of the M-discontinuity thus determined was about 10% greater than the results obtained from the earlier explosion studies. Kanamori (1963) considered that the seismic method, then, was not able to detect the existence of a second layer in the crust, thereby giving too shallow a depth for the M-discontinuity.

Aki (1961) and Kaminuma and Aki (1963) studied the crustal structure by the use of the phase velocity of Rayleigh waves with periods of 20–30 sec from distant earthquakes. Comparing the phase velocity of these waves with the standard phase velocity curves used by Press (1960), they tried to estimate the thickness of the crust in various regions in Japan. They discovered that the crustal thickness estimated from the Rayleigh waves is always too large in comparison with the thickness determined by the explosion studies, the difference being about 20 km. They showed further that the results can be put in harmony if the wave velocity of all the layers in Press' model (6EG) is reduced by 5.5% (Fig.21). They called this revised model 6EJ, and produced a map of the crustal thickness in Japan as shown in Fig.22 (Kaminuma and Aki, 1963; Kaminuma, 1964). These results were generally consistent with those from the explosion studies and also with Kanamori's (1963) results.

On the other hand, Aki and Kaminuma (1963) investigated the same problem by also analyzing the phase velocity of Love waves with the period of 30–36 sec. They

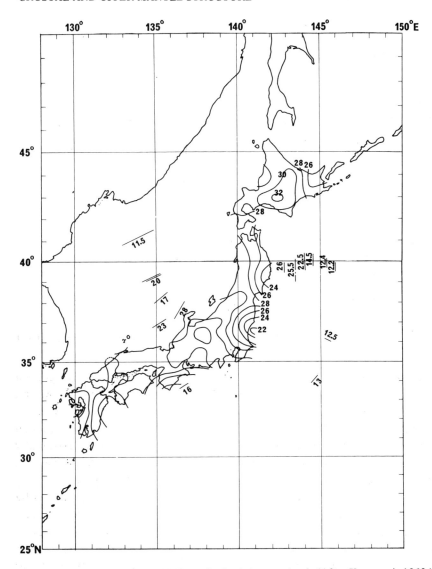

Fig.20. Depth in km of the M-discontinuity below sea-level. (After Kanamori, 1963.) Values in the oceanic areas are after S. Murauchi (personal communication, 1967).

discovered that the observed phase velocity dispersion did not agree with the phase velocity curves of Rayleigh waves based on 6EJ, i.e., the model good for Rayleigh waves was not good for Love waves. One possible explanation was that the upper layers under Japan are anisotropic in such a way that the velocity of the SH-wave is several percent greater in the horizontal direction than in the vertical direction (Kaminuma, 1966). In order to explain such an anisotropy, Shimozuru (1963) and Takeuchi et al.

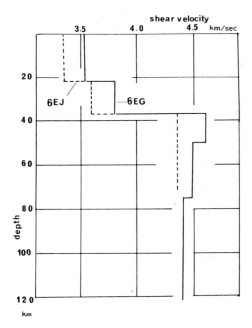

Fig.21. Shear velocity distribution model 6EG for a standard continental crust (Press, 1960), and 6EJ for Japan (Aki, 1961).

(1968) suggest that a portion of the upper mantle is molten and that the molten portion is dispersed in the upper mantle in the form of horizontal discs. Recently, however, Noponen (1969) examined the crustal structures of the area by using surface waves of longer periods (20–90 sec for Rayleigh waves and 20–65 sec for Love waves) and found that the discrepancy between Rayleigh wave and Love wave results does not exist under west Japan. Because of the difficulty inherent in the use of Love waves of shorter periods, there seems to be a possibility that the discrepancy cited above is not always real.

Mikumo (1966) also examined the crust models proposed by the explosion studies by comparing them with the gravity and surface wave data: using the reasonable velocity–density relations, the Bouguer anomaly and the phase velocity dispersion curves of Rayleigh waves were computed on a two-dimensional model along the eight explosion profiles in Japan (Fig.16), and comparisons were made with the observations. It was found that the single-layer crust models do not fit the gravity data, and two-or three-layered models, in which the existence of a lower crustal layer (Vp = 6.5–6.8 km/sec) and a 10–15 km thick intermediate layer (Vp = 7.4 km/sec) is assumed, were presented for central and western Japan: the results were essentially in harmony with Kanamori's. On the whole, the crustal thickness is more uniform in southwestern Japan, whereas in northeastern Japan the crust thins seawards quite regularly.

Making use of Rayleigh and Love wave dispersion data, Santo (1961, 1963) and Santo and Sato (1966) classified the world into fifteen regions. According to this study, the

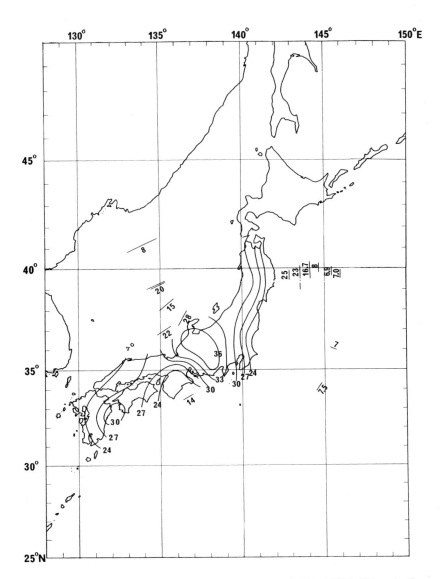

Fig.22. Crustal thickness in km. (After Kaminuma and Aki, 1963.) Values in the oceanic areas are after Murauchi (personal communication, 1967).

Japanese area belongs to region no.7. Saito and Takeuchi (1966) obtained probable upper mantle structures for various regions of the Pacific from Santo's dispersion curves. Their results indicate that the shear wave velocity immediately under the M-discontinuity for region no.7 is anomalously low (Vs = 4.3 km/sec). It was suggested that the upper mantle low velocity layer, which usually starts from a depth of 60 km or so, is raised under Japan up to the M-discontinuity. This agrees with the model of Aki (1961) and Kaminuma and Aki (1963) mentioned earlier. The upper mantle under Japan has been noted anomalous for its low Pn velocity for many years (Matuzawa et al., 1929; Jeffreys, 1952). This may be due to anomalously high temperatures in the upper mantle of active island arcs (see p.160).

If we examine the characteristics of the crust and upper mantle of the Japanese area more closely, some interesting features seem to emerge. The wave velocity in the upper layers under Japan seems to be highly heterogeneous. Recently, the Research Group for the Travel Time Curve (1968) produced a map of the distribution of station correction value (Cst) for shallow shocks (epicentral distance up to 1,500 km) as shown in Fig.23. Cst is the difference between the actual arrival time and the arrival time expected from the standard travel time curve, e.g., the Wadati–Sagisaka–Masuda curve (Wadati et al., 1933). Delay in the western side of the northeast Honsyu and Fossa Magna regions is clearly observed. This distribution is coincident with those of Cenozoic volcanism in the area.

From the anomalous distributions of earthquake intensity that have been known for many years, Katsumata (1960) and Utsu (1966) have shown that the Japanese Islands are underlain by an inclined low attenuating layer intruding from the Pacific side to the Sea of Japan side. Utsu (1967) and Utsu and Okada (1968) further investigated the wave velocity and attenuating properties of the region: it was concluded that, in order to explain the travel time anomalies of bodily waves from deep earthquakes in a consistent manner, the P and S wave velocities of the intruding low attenuating layer must be higher by about 6% than in the other portion of the upper mantle. The high-velocity zone is coincident in location with the deep-seismic zone under Japan but should be horizontally extended off the coast of Japan as shown in Fig.24. The existence of the high-velocity zone off Japan was previously shown by Hisamoto (1965). Kanamori and Abe (1968) have studied the velocity heterogeneity under the Japanese area using the surface waves from distant shocks. Their results confirmed the situations outlined above. By examining the travel times to Japanese stations from a nuclear explosion in the Aleutians, Kanamori (1968, 1970) concluded that P-waves travel slower by 0.4 km/sec in the upper mantle in the inner zone of the east Japan arcs than in its outer zone. It was further shown that the mantle above the deep seismic zone is so highly attenuating ($Q \sim 80$) that partial melting must occur. Situations strikingly similar to these remarkable arrangements, to which a great significance will be attached later (p.156 and 177), have also been disclosed for the Tonga Arc by Oliver and Isacks (1967). They call the intruding high-velocity, low attenuating layer the "lithosphere". We will use the same nomenclature on the following pages.

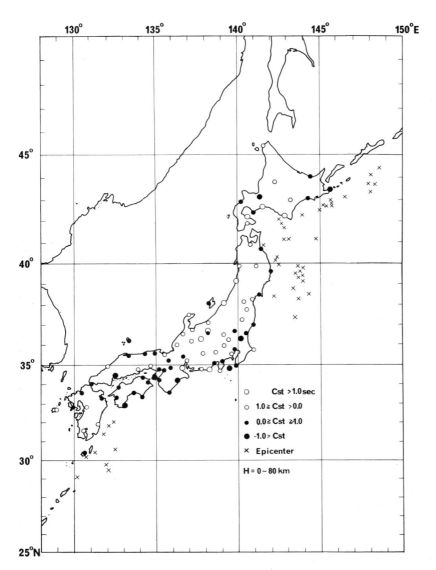

Fig.23. Distribution of station correction values of arrival time. (After the Research Group for the Travel Time Curve, 1968.)

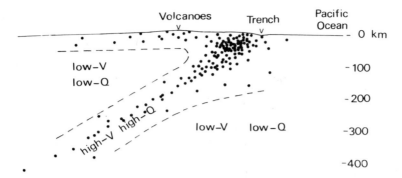

Fig.24. Upper mantle model under the Northeast Honsyu Arc deduced from seismic observations (Utsu and Okada, 1968). High $V = 8.2$ km/sec and low $V = 7.7$ km/sec.

SEISMICITY

As shown earlier in this book, most of the seismicity of the world is related to two principal environments: the mid-oceanic ridge and rift systems and the orogenic belt and island arc systems. Energywise, the latter systems, especially the island arc systems, are much more important. Moreover, the mantle earthquake zones are developed almost exclusively in the circum-Pacific island arc system. "Mantle earthquake zone" is used here for the zone under island arcs in which shocks emanate from depths ranging from 60 to 700 km.

In the rift—ridge systems, on the other hand there are generally only shallow earthquakes. The earthquakes associated with rift—ridge systems may be subdivided into two: namely those on the axis of the system and those on the "transform" faults cutting the system (Wilson, 1965; Sykes, 1967). The western part of the United States may belong to the latter, the San Andreas fault being one of the transform faults cutting the East Pacific Rise.

Japan and its environs show a typical island arc type. Fig.25 shows the regional distribution of the number of significant shallow earthquakes that occurred during 31 years from 1928 to 1958 in Japan and its environs with foci ranging in depth from 0 to 59 km. The epicenters of shallow earthquakes are roughly restricted in location to the area east of the Northeast Honsyu Arc and to the area of southwest Honsyu itself, the density of epicenters in the former area being much more pronounced.

As we have seen above, in the upper layers in the Pacific island arcs, great shallow earthquakes are occurring. The average rate of energy radiation by the earthquakes from the Japanese area (including the adjacent seas) is estimated to be $2.24 \cdot 10^{23}$ erg/year (Tsuboi, 1965). Divided by the area concerned, it gives the specific rate as $0.02 \cdot 10^{-6}$ cal/cm^2 sec. Since the seismicity is localized even within the Japanese area, as noted in the present section, the rate of energy discharge must also be localized, so that a value

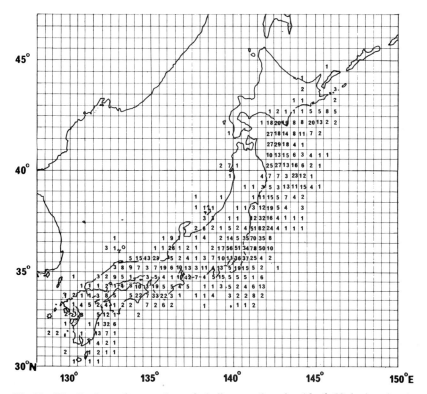

Fig.25. Distribution of a number of shallow earthquake ($d < 60$ km) epicenters in the period 1928–1958. (After Japan Meteorological Agency, 1958.)

much higher than $0.02 \cdot 10^{-6}$ cal/cm² sec, say, $01 \cdot 10^{-6}$ cal/cm² sec, would be more suitable for the active areas.

Fig.26 is the distribution map of the deep and intermediate earthquakes that occurred during the same 31-year period. Clearly the deep seismicity is related to the east Japan arcs and the southwestern portion (Ryukyu Arc) of the west Japan arcs. Southwest Honsyu is virtually free from deep shocks.

Generally, the deep and intermediate earthquakes in island arcs have an orderly arrangement in the depth of foci (e.g., Honda, 1934b; Wadati, 1935; Gutenberg and Richter, 1954; Sugimura, 1960). When the foci are plotted in a cross-section across an island arc, they generally lie around a plane that inclines downward from the oceanic trench toward the continent. This fact is well known for a number of arcs and the term "seismic plane" is used in this book for this inclined plane. Frequently, the zone is called the Benioff zone (Benioff, 1954). The existence of such a well-defined plane has been thoroughly shown by Sykes (1966) for other island arcs. In the case of the Tonga–Kermadec island arc, the hypocenters are so well determined in lying close to a "plane" that the thickness of the "mantle earthquake zone" is believed to be about 20 km or less

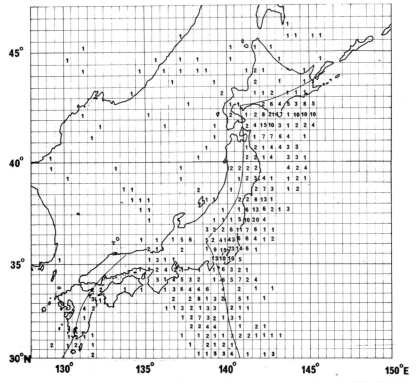

Fig.26. Distribution of a number of intermediate and deep earthquake ($d > 60$ km) epicenters in the period 1928–1958. (After Japan Meteorological Agency, 1958.) The thick line is the volcanic front (see p.55).

at some places. In the case of the arcs of Japan the planes appear to be curved in such a way as shown in Fig.27 by isobaths. The pattern also supports the view that the active island arcs in Japan extend from the Kuriles, via northeast Honsyu, to the Marianas but not to southwest Honsyu (see also p.15). Recently, Katsumata and Sykes (1969) have made detailed investigations on the seismicity of the Bonin–Marianas Arc and the Philippine Sea areas. It was found that the seismic plane under the Bonin–Marianas Arc is generally steeply inclined and, in particular, the plane is almost vertical at latitude 19° N. Making full use of data from nearby stations, Ishida (1970) also conducted a thorough examination of hypocenter distribution under the east Japan arcs. She established that the seismic zone under the east Japan arcs can also be so well defined that its thickness is the same as the limit of hypocentral determination, which is about 20 km.

EARTHQUAKE MECHANISM

Mechanism studies of the Japanese earthquakes by analysis of the initial motions have been made extensively by H. Honda and his collaborators (Honda, 1934a, b; Honda and

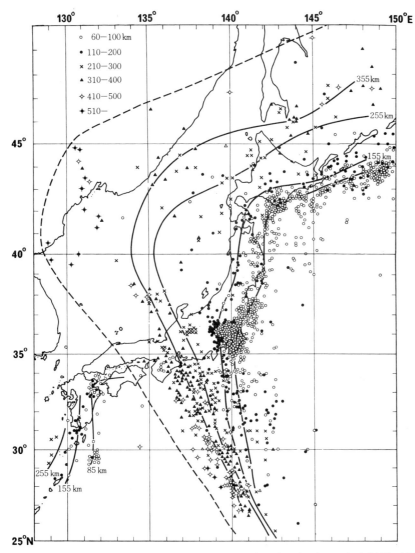

Fig.27. Epicenters of intermediate and deep earthquakes in the period 1928–1962. (After Japan Meteorological Agency, 1958, 1966.)

Masatsuka, 1952; Honda et al., 1957, 1967) and Ichikawa (1961, 1966, 1969, 1971) as shown in Fig.28 and 29. These studies made the pioneer contributions to the establishment of the double-couple mechanism of earthquakes (see, e.g., Hodgson, 1962). Recently, Ichikawa (1970) discussed the relation of the shallow shock mechanisms with the structure of arcs. They have also shown that in the Japanese area, the direction of the horizontal component of the maximum compression for the deep and intermediate earth-

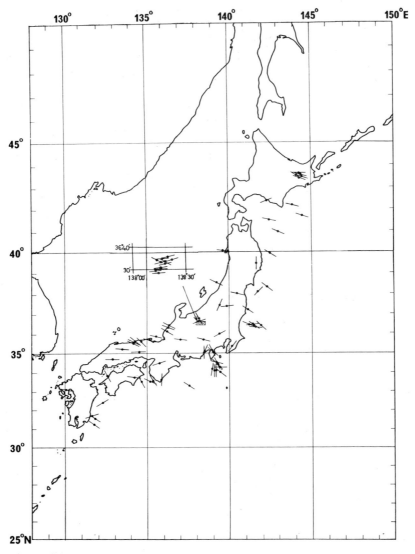

Fig.28. Directions of the maximum horizontal compressional stress for shallow earthquakes ($d < 60$ km). (After Honda, 1934a, b; Honda et al., 1967; Ichikawa, 1966, 1969, 1970.)

quakes is almost perpendicular to the direction of the arc as shown in Fig.29, indicating the existence of regional pressure exerted from the ocean toward the continent. This tendency is generally found in other circum-Pacific arcs (see Hodgson 1962; and p.86). For shallower shocks, the ordering in the direction of compressive axis seems to be different from that for the deeper shocks: the compressive axis lie mostly in the west–east and west-northwest–east-southeast directions. The areas of junction of Izu–Mariana

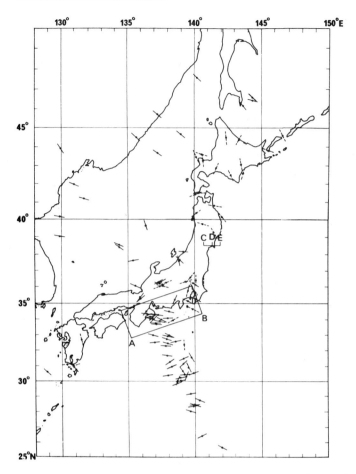

Fig.29. Directions of the maximum horizontal compressional stress for intermediate and deep earthquakes ($d > 60$ km). For C-D-E, see p.196. (After Honda et al., 1934–1967; Ichikawa, 1966.)

and the Southwest Honsyu Arc and of the Ryukyu and Southwest Honsyu arcs are exceptions in that the direction of the compressive axis is largely north–south (Ichikawa, 1969, 1970). Recently Stauder (1968) examined the direction of the after-shock forces of Rat Island earthquake in the Aleutians. He finds that the forces are tensional for the shallow shocks occurring outside the trench, and compressional for deeper shocks occurring inside. Shocks in Fig.28 are considered to be equivalent to those occurring inside the trench in Stauder's case. Earthquake mechanism study is extremely important in examining the forces and movements causing the earthquakes, which in turn may be the present manifestation of the forces and movements governing tectonics. In the frame work of the new global tectonics (see, e.g., Isacks et al., 1968), these forces and movements represent the relative motions between adjoining plates. McKenzie and Parker

(1967), in their first paper proposing the plate-tectonics, showed that the crust and uppermost mantle of the north Pacific behave like a rigid plate.

Now, as seen in Fig.30, the direction of compressional axis in the vertical plane is predominantly parallel to the "seismic plane" at least in the Izu–Mariana Arc region. It is remarkable that this tendency is held well when the dip of seismic plane varies (Katsumata and Sykes, 1969). It thus appears that the "seismic plane" is not the plane of maximum shearing stress, but it is oblique to the plane of maximum shear or the fault plane by some 45°. If we choose, from the two theoretically possible fault planes, the vertical focal slip plane, it may be inferred that the slips are characterized by the rise of the western side and the drop of the eastern side. Aki (1966) investigated the average focal stress patterns by determining a single solution for a group of earthquakes occurring in 1961–1963: sixteen areas in Japan were selected for this purpose and average stress patterns for each area were obtained. The results were in almost perfect accordance with the studies by H. Honda and others. A similar tendency has been found for the Tonga–Kermadec Arc by Isacks et al. (1969). Possible interpretations of these observations will be given in Chapter 3 (p.190). It will be shown there that the features described here may be understood as being due to the sinking of the lithospheric plate along the seismic plane.

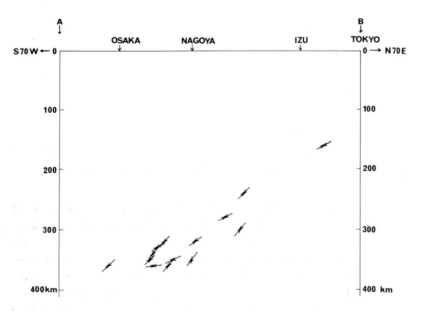

Fig.30. Directions of the principal compressional stress for intermediate and deep earthquakes in a vertical plane. (Data from Honda et al., 1967.) Arrows are the projections on the vertical plane passing AB in Fig.29.

SURFACE FAULTING

Deformations observed in the epicentral region of major earthquakes, in particular the surface earthquake faults, are the most remarkable type of crustal deformations. An earthquake fault is referred to as a fault exposed on the surface, accompanying an earthquake (Otuka, 1936; Iida, 1959). Table II and III and Fig.31 present the information on recorded earthquake faultings in Japan. The investigation of a number of recorded earthquake faults has established that many of them coincide in position with the faults inferred from geological and geomorphological evidence especially with zones of recent faulting inferred from the tectonic relief, including a fault valley, a series of saddles, and offset of a river system. Major earthquakes cause many cracks in the ground along banks, roads, and other features. They are the phenomena restricted to only the shallowest part of the crust, and are not usually called earthquake faults. An earthquake fault manifests itself for a longer distance, crossing mountains and valleys. They are suspected to reach deeper parts of the crust.

TABLE II

Historical major earthquakes accompanied by surface faults in Japan

No.	Name	Date	Magnitude[1]
1	Nôbi (Mino–Owari)	28/10/1891	8.0
2	Sakata (Syônai)	22/10/1894	7.3
3	Rikuu (Ugo)	31/ 8/1896	7.5
4	Anegawa (Kônô)	14/ 8/1909	6.9
5	Kanto (Sagami Bay)	1/ 9/1923	7.9
6	North Tazima	23/ 5/1925	7.0
7	Tango (Okutango)	7/ 3/1927	7.4
8	North Izu	26/11/1930	7.0
9	Tottori	10/ 9/1943	7.3
10	Mikawa (Atumi)	13/ 1/1945	6.9
11	Hukui	28/ 6/1948	7.2
12	Hutatui	19/10/1955	5.7
13	Niigata	16/ 6/1964	7.3
14	Matusiro earthquake swarm	Apr.–Sept./1966	Max. = 5.0

[1] According to the Richter – Gutenberg scale.

Putting aside some of the volcanic earthquakes most shocks generated in the crust are believed to be vibrations caused by shear failure of rocks. Focal faulting is suspected for every occurrence of earthquake, and if the focal fault extends to the surface the surface fault is defined as primary. Horizontal deformation of the ground around a few of the primary faults was measured by means of triangulation and was revealed to coincide well with horizontal displacement of the faults. The character of displacement of the primary

TABLE III

Historical earthquake surface faults in Japan

No.	Name of fault	Type	Maximum displacement	
			vertical (m)	horizontal (m)
1−1	Neo Valley	A_1	2.0	7.2
1−2	Kurotu−Nukumi	A_1	1.8	2.7
1−3	Inazawa	C	0.3	n.d.
1−4	Midori	A_1	5.5	2.0
2−1	Yadarezawa	A_2	n.d.	−
3−1	Kawahune	A_2	2.0	−
3−2	Sen'ya	A_2	2.5	−
4−1	Yanagase	C	0.2	n.d.
5−1	Sitaura (Nagasawa)	B	1.5	−
5−2	Sinkawa	B	1.0	−
5−3	Enmeizi	B	1.0	−
6−1	Tai	B	0.5	−
7−1	Gômura	A_1	0.6	3.0
7−2	Yamada	A_1	0.7	0.8
8−1	Tanna	A_1	2.0	3.5
8−2	Himenoyu	A_1	0.3	1.0
9−1	Sikano	A_1	0.8	1.5
9−2	Yosioka	A_1	0.5	0.9
10−1	Hukôzu (Katahara)	A_2	2.0	1.0
10−2	Yokosuka (Ehara)	B	1.2	0.6
11−1	Hukui	C	0.7	1.0
12−1	Hutatui	C	0.1	n.d.
13−1	S_3	B	5	−
13−2	Awasima−Utiura	B	0.7	−
14−1	Matusiro	A_1	0.3	0.3
14−2	F14	B	0.3	−

− = zero or very small.

faults is generally in harmony with the direction of compressions and dilatations of the main shocks and the associated after-shocks. There is another group of earthquake faults, the secondary faults. The secondary faults differ from the primary ones in showing a direction of displacement which does not harmonize with the regional deformation accompanying the earthquake. Many of them show no harmonic relationship with the radiation pattern of the associated earthquakes either. Matsuda (1967) calls the primary and secondary earthquake surface faultings A-type and B-type (Table III).

In general, the displacement of a primary fault consists both of a horizontal component and a vertical component with various ratios. Depending on the predominating component, the fault is further classified into a primary strike-slip fault (A_1 in Table III) and a primary dip-slip fault (A_2 in Table III). In Japan, the Gomura fault from the Tango earthquake (1935), the Tanna fault from the North Izu earthquake (1930), and the Sikano fault from the Tottori earthquake (1943) are famous as the primary

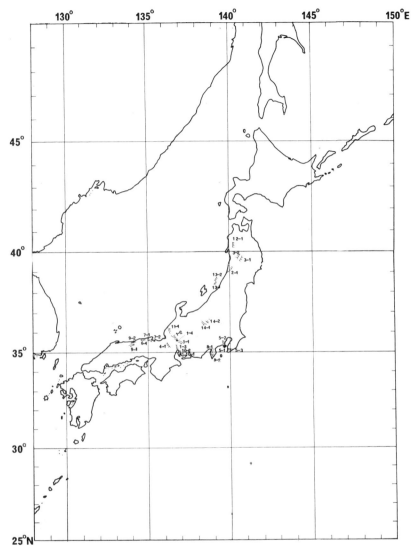

Fig.31. Distribution of earthquake faults. Numerals refer to Table III. The Midori fault is indicated by a cross.

strike-slip faults (see Table II, III; Fig.31). The secondary faults are believed to be caused by the regional deformation accompanying the earthquakes or other indirect causes. The Kanto earthquake (1923) provided several examples of the secondary faults, which show dip-slip displacements within the broadly uplifted epicentral area. Most portions of the Neo Valley fault, along which a series of fault scarps formed for 50 km in 1891, are also the representative primary strike-slip faults (Fig.32). Midori fault scarp is a part of the

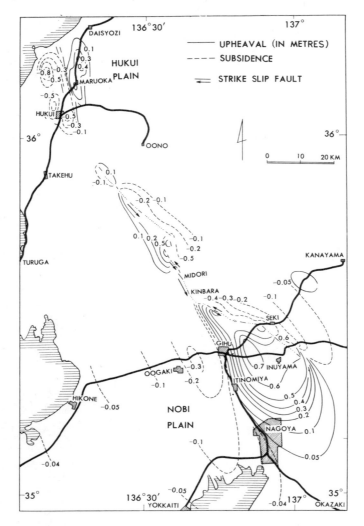

Fig.32. Vertical displacements caused by the Mino–Owari earthquake (1891) and the Hukui earthquake (1948). (After Muramatu et al., 1964.)

Neo Valley fault system and shows a displacement of about 5.5 m vertically and about 2 m horizontally. The amount of the vertical displacement differs from ordinary ones known for other parts of the Neo Valley fault system as well as for other strike-slip earthquake faults, but it was revealed by the examination of the underground structure by Muramatu et al. (1964) that the Midori scarp is not a subordinate surface feature but a part of a complicated assemblage of faults in the Paleozoic basement rocks.

In this book C-type earthquake faults may be added to Matsuda's list. They are the faults not observed on the surface but confirmed by releveling or other means, and are called geodetic faults. Exactly speaking, the result of the releveling comprises the deformation before and after the earthquake, and the displacement that is considered to be too large to have occurred only during quiescent periods, is interpreted as accompanying the earthquake. However, some of them may not be faults but flexures.

The distribution of earthquake faults in Japan (Fig.31) covers the area where volcanic activity is intensive (Fig.39, see p.57) and its outskirts. But, the positive correlation in regional distribution between volcanoes and earthquake faults does not mean local coincidence between them. No earthquake faults are known to cut a volcanic cone except Tanna fault (8-1 in Table III; Fig.31) that extends into the Hakone Volcano. It is interesting to note the fact that in the Late Cenozoic sediments and sedimentary rocks, such as in the Uetu zone (12-1, 13-1 and 13-2 in Table III), the Kanto Plain (5-1, 5-2 and 5-3 in Table III), the Nobi Plain (1-3 and 10-2 in Table III), and the Hukui Plain (11-1 in Table III), only the secondary (B-type) and geodetic (C-type) faults have been observed, whereas in the basement terrain of volcanic and granitic rocks such as in north Izu, Mikawa, Neo Valley, Tango and Tottori, the primary (A-type) faults have been observed. Apparently, the focal faulting cannot reach the surface through a thick layer of soft material whereas it can do so through hard materials.

As to the primary (A-type) earthquake faults in Japan, strike-slip faults and reverse faults are known while normal faults are not known. This fact is in harmony with the tendency of the radiation pattern of very shallow Japanese shocks in which the normal fault type pattern is scarcely found (Ichikawa, 1966). As seen in Fig.31, three of four dip-slip reverse faultings occur in northeast Japan, while strike-slip ones predominate in central Japan. The area where these strike-slip faults cluster corresponds to the area of the junction between the east Japan island arcs and west Japan island arcs. Strike-slip surface faults associated with five major earthquakes (1-1, 1-2, 7-1, 7-2, 8-1, 8-2, 9-2 and 11-1 in Table III and Fig.31) and one earthquake swarm in 1966 (14-1 in Fig.31; see Nakamura and Tsuneishi, 1967) form a broad system of conjugate strike-slip structures, one trending northerly or northwesterly and another easterly or northeasterly (Sugimura and Matsuda, 1965).

It may be added here that Ludwig et al. (1966) found, by seismic reflection technique, numerous "normal" faults extending to the *surface* (see Fig.95, p.141) on the oceanward wall of the Japan Trench. These may be of recent origin. Although the strikes of these faults are unknown and therefore the possibility of other types of faults than the normal ones remains, this seems to correspond to the tensional stress patterns found for earthquakes in the area oceanward of the Aleutian Trench (Stauder, 1968; see also p.39). In the framework of plate tectonics, such tensional forces are due to forced downward bending of the oceanic plate (Isacks et al., 1968).

CRUSTAL DEFORMATION

In most parts of Japan, the earth movements during the last several decades have been determined precisely by means of resurveying of the leveling routes and the triangulation nets. Land surveying in Japan is carried forward at present by the Geographical Survey Institute of the Ministry of Construction. The start of this work was as early as 1871. Up to 1945 the surveying was worked out by the Land Survey Department of the Army. When earth movements took place suddenly during disastrous earthquakes, resurveying was made in the area concerned. As a result, much valuable data on horizontal and vertical movements accompanying earthquakes have been obtained. Resurveying the areas where no major earthquake occurred has made clear that the earth movements are taking place during quiescent times also, but at a much smaller rate. Observation of crustal deformations has been taken up as one of the most important projects in the Japanese program of earthquake prediction (Hagiwara and Rikitake, 1967).

It is well known that land has risen and fallen even during recent years in such restless belts of the crust as that of Japan. The fact has been detected mainly by releveling of bench marks, although other means have also contributed to reveal the earth movements. A first-order level net, which consists of leveling routes, about 20,000 km in total length, covers all of Japan. The bench marks for leveling survey were set up at intervals of about 2 km along main national roads and about 1 km along new highways. Their heights are measured repeatedly: regular levelings were carried out three times, in 1894–1906, 1922–1934 and 1948–1957. Incidental local resurveys of first-order leveling have been made after disastrous earthquakes and volcanic eruptions. They showed considerable changes in level.

Miyabe (1952) calculated secular vertical movement along the leveling network throughout the main part of Japan by adjusting the data of the 1894–1906 and 1922–1934 regular levelings to a continuous consistent change from route to route with a common period from 1900 to 1928. He distributed the errors to every measurement along a route by proportional allotment. Although it cannot be helped making such adjustments in compiling the data of different periods, the results may be regarded as representing, at least approximately, the true vertical movements. Miyabe et al. (1966) illustrated the result as a map as reproduced in Fig.33 in a simplified form. The rectangles in this figure mark the areas showing deformations affected directly by major earthquakes and volcanic eruptions. These deformations and the results of other recent research will be discussed later in this section.

The whole area of Japan is fairly densely covered by a network of triangulation points. The Geographical Survey Institute of Japan is repeating geodetic surveys through the first-order triangulation nets. The systematic resurvey was carried out until 1963, and the special resurveys were made over limited areas after major earthquakes, during the period of the Matusiro earthquake swarm, and during the years of the Upper Mantle

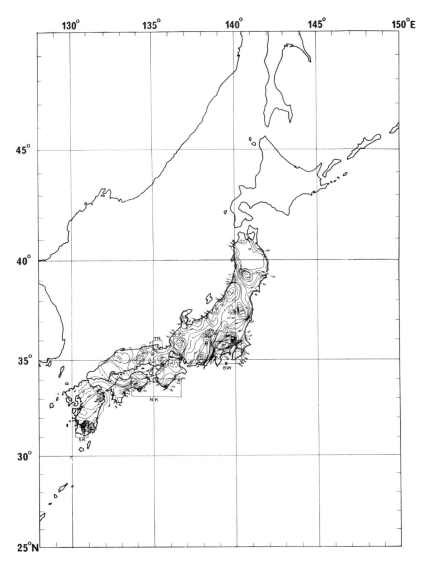

Fig.33. Vertical displacement detected by precise releveling. (After Miyabe et al., 1966.) *KW, NK* and *TN* are the epicentral areas of major earthquakes; *SK* is the subsiding area around the Sakurazima Volcano. Contour interval = 2 cm/28 year (1900–1928).

Project. Their surveys were supplemented by geodimeter studies with more frequent repetition of measurements.

A first-order triangulation with about 300 points was first carried out in 1883–1909, and re-triangulation was completed over the same first-order nets in 1948–1967. Comparison of these data makes it possible to give the pattern of horizontal surface

Fig.34. Horizontal displacement detected by precise re-triangulation and net-adjustment of multiple fixed stations. (After Harada and Isawa, 1969.)

deformation of the Japanese Islands during the past several decades. Harada (1966, 1967) and Harada and Isawa (1969) compiled the results of re-triangulation over the whole country. The essential part of the map of Harada and Isawa of the distribution of vectors of horizontal displacement of first-order triangulation points is shown in Fig.34. Significant horizontal displacement is taking place in the areas that experienced major earthquakes during the period between the two surveys.

A comparison of the mean sea-level data obtained at multiple mareographs along the Japanese coasts enables us to know relative ground movements with a certain precision (Tsumura, 1964). Mareographs are established and controlled as follows: thirty-two by the Meteorological Agency, nine by the Geographical Survey Institute, five by the Maritime Safety Board, and ten by other agencies.

The water-tube tiltmeters and the extensometers are used to study the slight crustal deformations. These instruments are useful in recording the continuous changes in the local inclination of the earth's surface and in the length of a part of the ground during quiescent periods. Information obtained by such station measurements constitutes an important complement to that from leveling and triangulation resurveys, which are regional but conducted, in practice, only intermittently. Tiltmeters and extensometers have been operated at more than a score of localities scattered over the Japanese Islands by the Earthquake Research Institute and other university institutes. Results of these observations are to be found in, e.g., Ichinohe and Tanaka (1964, 1966).

The most remarkable expressions among the movements are the seismic uplifts and subsidences (*TN, NK* and *KW*, in Fig.33) and the movements accompanying volcanic eruptions (*SK* in Fig.33). The subsidence in the coastal area around Kagosima Bay (*SK*) at the time of the explosion of the Sakurazima Volcano in 1914 is a typical example showing a close relationship with the volcanic activity (Mogi, 1958).

At the time of the Nankai earthquake in 1946 (*NK*), an uplift of tens of centimetres to one metre was observed in the southern tip of the Kii Peninsula (see Fig.63 on p.92 for location) and Muroto Cape and Asizuri Cape in Sikoku, and a subsidence of a little less than one metre occurred in the root areas of these peninsulas where sea water invaded into the lowlands. Also, at the time of the Kanto earthquake in 1923 (*KW*), the southern part of the Miura Peninsula and Boso Peninsula were uplifted and the inland areas including Tokyo were subsided. At Muroto Cape, as discussed by many authors, the history of sudden uplifts associated with destructive earthquakes such as the one in 1946 and chronic subsidences between two earthquakes are documented for a few centuries (e.g., Sassa, 1951). The amount of net uplifts a, that is, the sudden one minus the chronic subsidence during the period between two destructive earthquakes, is about one fifth of the amount of one sudden uplift. The total amount A of the net uplift for the last 6,000 years can be assessed from the heights of the Holocene highest terrace and of Holocene sea-level (Sugimura and Naruse, 1954). A/a is about 55 for Muroto and Kanto areas. This might mean that about 55 destructive earthquakes have taken place in these areas during the last 6,000 years. This is consistent with the historical average interval of major earthquakes, 110 years in Muroto and 140 years in Kanto. The deformation in the Muroto area can be traced back to Middle Pleistocene with a constant rate of uplift (see p.133). As exemplified by the earth movements in the Muroto and Kanto areas, there

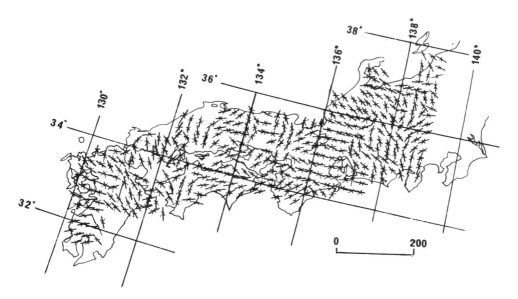

Fig.35. Direction of compressional axis of horizontal strain. (After Kasahara and Sugimura, 1964b.)

seem to be pulses with intervals ranging from several decades to several centuries. As a result, the authors consider that the map of earth movements during the last several decades may not necessarily represent the geologically significant trends because of the shortness of the time.

Dambara and Hirobe (1964), Miyamura and Mizoue (1964), Dambara (1968), Mizoue (1968) and others also published summaries of earth movements detected by means of releveling and by re-triangulation in some areas of disastrous earthquakes.

Kasahara and Sugimura (1964a, b) used the data from the triangulation stations in central and western Japan and computed the strain components, such as dilatation, maximum shear, and principal axes. Fig.35 shows one of the results, i.e., the distribution pattern of the directions of maximum principal strain, which is similar to that of Late Quaternary strike-slip faults in central Japan. The regional tectonic uniformity that may be expressed by the persistent directions of axes of horizontal principal stress deduced from the faults is well illustrated in this region (Matsuda, 1967). Dip-slip varieties are also included in the Late Quaternary and recent primary faulting in central Japan. But the vertical displacements have been small in general (Table III), so that the horizontal ones seem at present to give the essential features of the earth movements.

GEOLOGICAL STRUCTURE

The main part of the Japanese Islands is, for the convenience of description, divided by a line provisionally called the *Sapporo—Tomakomai Line* and a fault called the *Itoigawa—Sizuoka Line* (both in Fig.37, see p.54) into three parts: *Hokkaido, northeast Japan* and *southwest Japan.*

The western zone (*IA* and *IB* in Fig.36) of northeast Japan is largely made up of folded Late Cenozoic sediments overlain by Quaternary volcanic rocks, and underlain by Mesozoic and older rocks which outcrop only in places between them. The southwestern peninsula of Hokkaido is the northern continuation of this western zone. The southern continuation of it is the intensively deformed area east of the Itoigawa—Sizuoka Line. This area is customarily called *Fossa Magna* (see p.113). The Itoigawa—Sizuoka fault crosses and cuts the structure of southwest Japan, which reappears beyond Fossa Magna in Kanto Range, a part of northeast Japan. The eastern zone of northeast Japan consists of the Kitakami Plateau, the Abukuma Plateau, the Kanto Range, and some other ranges. Paleogene and older rocks are distributed in these areas (see Fig.63, p.92, for the names of places cited).

Southwest Japan is divided by the *Median Tectonic Line* (Median Dislocation Line or simply Median Line), which trends almost east—west, into two zones, i.e., the *Outer Zone* and the *Inner Zone.* The Median Tectonic Line has long been thought to be a thrust with which the Inner Zone overlies the Outer Zone, but the need of reexamination is arising because it was found recently to be a right-lateral fault at least during the Quaternary Period (Kaneko, 1966; Okada, 1968). The western extension of this fault to

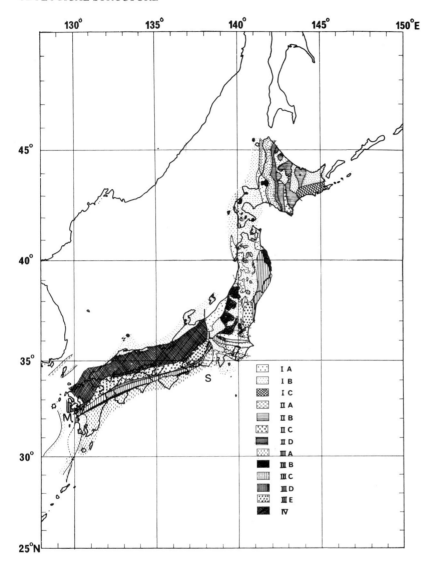

Fig.36. Tectonic map. (After the Geological Survey of Japan, 1968.) *I A* = Late Cenozoic fold zone, structural basin, and sedimentary basin; *I B* = Middle Cenozoic volcanic belt; *I C* = Early Cenozoic and latest Mesozoic sedimentary basin; *II A* = Fold zone of Alpine-age orogeny; *II B* = Low-temperature metamorphic and plutonic belt of the Alpine-age orogeny; *II C* = High-temperature metamorphic and plutonic belt of the Alpine-age orogeny; *II D* = Unmetamorphosed zone of the Alpine-age orogeny; *III A* = Early Cenozoic and Middle and Late Mesozoic geosyncline; *III B* = Early Mesozoic geosyncline; *III C* = Late Paleozoic geosyncline (unmetamorphosed zone); *III D* = Mesozoic low-temperature metamorphic belt; *III E* = Mesozoic high-temperature metamorphic and plutonic belt; *IV* = Paleozoic orogenic belt. *I—S* = Itoigawa—Shizuoka Line; *M—L* = Median Tectonic Line.

Kyusyu is made obscure by the cover of Cenozoic sediments and volcanic rocks. The eastern extension of the fault will be mentioned later (see p.113).

The Inner Zone (*IV* and *IIIE* in Fig.36) of southwest Japan consists of Paleozoic and Mesozoic sedimentary and metamorphic rocks as well as plutonic rocks intruding them, covered by Cenozoic sediments. In this zone, granitic batholiths and acid volcanic pyroclastic flows predominate. In the Outer Zone, there is an east—west trending zonal arrangement of tectonic provinces, which are as follows from north to south:

(*1*) Sanbagawa metamorphic belt, including Sanbagawa crystalline schist and Mikabu sheared rocks (*IIID* in Fig.36).

(*2*) Titibu (Chichibu) unmetamorphosed belt, including Paleozoic and Mesozoic sedimentary rocks with a tectonic horst characterized by the metamorphic rocks of pre-Silurian age (*IIIC* and *IIIB* in Fig.36).

(*3*) Simanto belt, which consists largely of non-fossiliferous and partly fossiliferous Mesozoic and Paleogene sediments overlain unconformably by Neogene sediments (*IIIA* in Fig.36).

The Outer Zone is characterized by a small amount of intrusive rocks.

The Nansei Syoto Islands constitute the Ryukyu Arc which has essentially the similar history as the Outer Zone of southwest Japan. But the Ryukyu Arc is characterized by its contemporary island arc activities.

Hokkaido contains a belt of metamorphic rocks (*IIB* and *IIC* in Fig.36) in its middle area in a northerly direction pointing Sakhalin. A belt of deformed Jurassic strata with granitic and ultrabasic intrusives makes up the main part of the belt. A fold belt of Cretaceous, Paleogene and Miocene strata (*IIA* in Fig.36) is on the west next to the north—south trending metamorphic belt.

The Izu Seven Islands (Izu—Sitito Islands, see Fig.8) are a Cenozoic volcanic arc. The Ogasawara or Bonin Islands have a Paleogene and later history. Limestones and a volcanic assemblage make up the Ogasawara main islands, whereas the Kazan or Volcano Islands in the southwest consist of Quaternary volcanoes.

Downfaulted or downwarped basins, scattered over the Japanese Islands, trap Quaternary System, which forms plains and neighbouring terraces and hills. Kanto Plain is the largest of these basins (see p.129).

The Geological Survey of Japan (Isomi, 1968) has recently published tectonic maps of Japan. The maps were prepared for the international cooperation on the world tectonic map and were presented before the 23rd International Geological Congress in Prague in 1968. The idea of preparing the world tectonic map was originated from the conception of "structural stage" (Bogdanoff, 1963). In Japan, the whole area was broadly divided into four orogenic terrains and each terrain in turn was subdivided into a few zones according to the structural stages in each orogeny. Here we reproduce one of the tectonic maps with some abbreviations including omission of the subdivisions of the oldest orogenic terrain (Fig.36). In Fig.36 we have to notify that the Late Cenozoic active zone (*IA* and *IB*) extends along the inner side of the east Japan island arcs (see Fig.10), showing remarkable parallelism to the island arc features (see also Fig.80, p.124). This subject will be more fully discussed in Chapter 2.

A compilation of the distribution of geologically observed faults in Japan is provided by Murai (1966) and the distribution of some folded structures in Japan is discussed by Kimura (1968). Murai shows a map of all faults collected from the so far printed maps without any subjective view.

We must now mention of pairs of metamorphic belts (Hashimoto et al., 1970). There are pairs of two different series of metamorphic facies lying parallel in the circum-Pacific orogenic belts inclusive of the Japanese Islands, New Zealand, Celebes, and even the western North America. The last named area is not an island arc type region at present, but seems to have been an island arc in the geological past (Hamilton, 1969). In these circum-Pacific regions, a metamorphic belt of the jadeite—glaucophane type or high-pressure intermediate group extends on the Pacific side, while a belt of the andalusite—sillimanite type or low-pressure intermediate group on the continental side (Miyashiro, 1961). Between these belts, in most cases, there is a major fault. Generally high p/t metamorphic belts are accompanied by ultrabasic rocks and low p/t belts are accompanied by abundant granitic rocks. It is generally considered that the associated two belts were formed probably by one and the same orogeny. These orogenies in the circum-Pacific regions are older than the formation of the present island arcs. It will be later discussed why a pair of a high p/t type and a low p/t type exists, in connection with the processes under island arcs (see p.200).

As shown in Fig.37, three pairs of metamorphic belts are exposed in the Japanese Islands (Miyashiro, 1961). The youngest pair is composed of Hidaka and Kamuikotan metamorphic belts, lying parallel to each other. The metamorphism of these belts is a part of the Hidaka orogeny. The metamorphism that is related to the next older orogeny in Japan took place along the Pacific side of Honsyu. The pair is composed of Ryoke—Abukuma or simply Ryoke metamorphic belt and Sanbagawa metamorphic belt, lying parallel to each other. The regional metamorphism of Sanbagawa belt is glaucophanitic, producing glaucophane, lawsonite and jadeite. This metamorphism took place under a higher pressure and a lower temperature than that of Ryoke belt. The Sanbagawa belt belongs to the metamorphic belt of jadeite—glaucophane type and high-pressure inter-mediate group. On the other hand, the regional metamorphism in the Ryoke terrain produced andalusite, cordierite, and sillimanite in rocks of appropriate compositions and are accompanied by a group of batholiths of granitic composition. The Ryoke belt is characterized by a high-temperature and low-pressure type metamorphism, and belongs to the metamorphic belt of andalusite—sillimanite type and low-pressure intermediate group. The oldest pair consists of Hida and Sangun metamorphic belts, lying parallel to each other along the Sea of Japan side of southwest Japan. The Hida belt is assigned partly as the andalusite—sillimanite type and partly as the low-pressure intermediate group. Most parts of the Sangun belt seem to belong to the high-pressure intermediate group metamorphism.

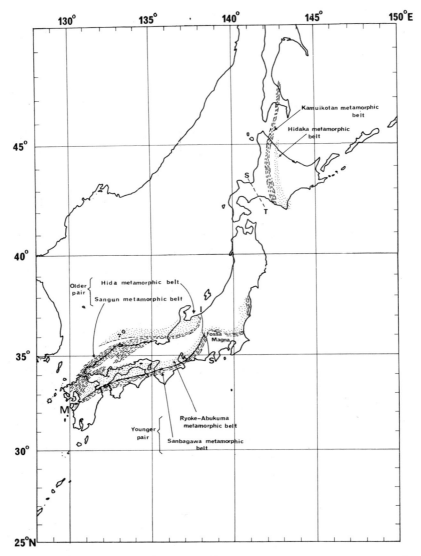

Fig.37. Metamorphic belts. Area shaded with dots: metamorphic belt of low-p/t type; area shaded with short lines: metamorphic belt of high-p/t type. (After Miyashiro, 1961a.) The line $I–S$ indicates the Itoigawa–Shizuoka Line, the line M the Median Tectonic Line and the line $S–T$ the Sapporo–Tomakomai Line.

These three parts correspond to the three orogenic belts represented by the numbers *II, III,* and *IV* from younger to older in Fig.36.

Radiometric age datings are now available for igneous and metamorphic events. For example, Kawano and Ueda (1966) dated igneous rocks by the K–A method, Hayase and Ishizaka (1967) showed ages by the Rb–Sr method, and Ozima et al. (1967) investigated by both the K—A and Rb—Sr methods. Summaries of available ages until 1969 are provided by Nozawa (1970) for Late Cretaceous acidic igneous rocks and by Kaneoka and Ozima (1970) for volcanic rocks. Correlation with the Cenozoic biostratigraphical consequences was tabulated by Chinzei (1967) and Ikebe and Chiji (1969).

VOLCANOES

Fig.38 shows the distribution of the Quaternary volcanoes in the area. This distribution also witnesses the validity of the notion of the east Japan and west Japan arcs. Active volcanoes are distributed in two distinct belts: the one starting from the Kurile, running through northeastern Honsyu, and bending southward to the Izu (Sitito)— Mariana Arc; the other from Kyusyu to the Ryukyu Arc. These two belts will be called the east Japan volcanic belt and the west Japan volcanic belt, respectively. In the past, it was considered that there were many more volcanic belts in the area. But, due to the advances in volcano-stratigraphy, some volcanoes have been identified as extinct so that the distribution of active volcanoes has become considerably simpler. We believe, at the present stage, that the above two belts are genetically significant, and further subdivisions of them into many small belts are of no use.

The trend of the two belts is clearly parallel to the trend of the oceanic trenches (Fig.9). In the southwestern Honsyu, i.e., the area bounded by the two active belts, both volcanic belt and trench (the Nankai Trough, see p.16) are much less pronounced. Another important fact to be observed in Fig.38 is that the volcanoes exist exclusively in zones whose boundaries facing the ocean is no less than 200 km away from the trench axis. Those oceanward boundaries of the volcanic zone were named the "fronts of the volcanic belts" or "volcanic fronts" by Sugimura (1960). It also is noteworthy that the population density of volcanoes is the greatest just continentward of the front and decreases with distance from it toward the continent. Such an asymmetric distribution of volcanoes is common to other island-arc areas (Suzuki, 1966). We believe that the existence of a clear front and the asymmetric distribution of active volcanoes have a definite genetic significance in the tectonics of island arcs (see pp.156—177).

Sugimura et al. (1963) made a survey of volumes of Quaternary volcanic products in Japan and compiled the results as shown in Fig.39. In estimating the volume of a volcano, these authors substituted a volcano by a cone having the same basal area S as that of the volcano and a height T that is equal to the difference in altitude of the summit and the highest exposure of the basement. They also made use of an empirical formula:

$$T = 43.5 \times S^{0.5} \text{ or } V = 0.0145 \times S^{1.5} \tag{4}$$

thereby making it possible to estimate the volume V of a volcano by measuring its S only. Their results, for instance, say that the volume of Mount Huzi (Fuji) is 390 km^3.

It seems to be an established fact that the major fraction of energy of volcanic eruption is transported in the form of thermal energy of the ejecta (e.g., Verhoogen,

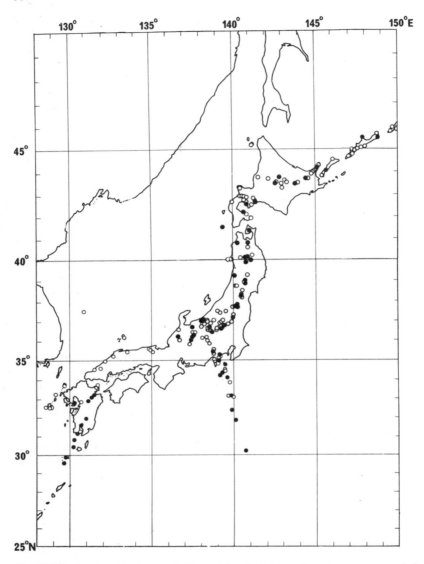

Fig.38. Distribution of volcanoes in Japan. Closed circles: active volcanoes; open circles: other Quaternary volcanoes.

1946; Sugimura, 1968). It has been shown that the average rate of energy output by volcanism is $0.06 \cdot 10^{-6}$ cal/cm^2 sec for the Izu volcanic islands south of Tokyo for the last 10^6 years (Yokoyama, 1956, 1957a, b), and $0.04 \cdot 10^{-6}$ cal/cm^2 sec for the same period in the whole volcanic zones of Japan (Sugimura et al., 1963). A value of $0.3 \cdot 10^{-6}$ cal/cm^2 sec has been obtained for the last 50 years for the Japanese volcanic zone by the same authors. These values compare the rate of seismic energy output of $0.02-0.1 \cdot 10^{-6}$

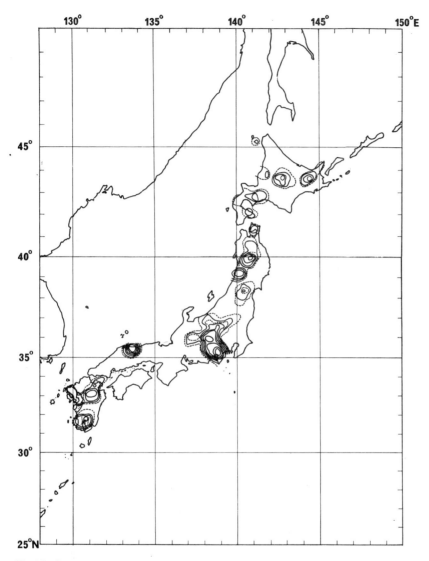

Fig.39. Contours showing the quantitative distribution of the Quaternary volcanic materials in thickness. Contour interval: 20 m (solid line) and 10 m (broken line). (After Sugimura et al., 1963.)

cal/cm^2 sec in Japan and environs. It must be noted that much higher values of volcanic energy have been reported for other volcanic areas: $1.8 \cdot 10^{-6}$ cal/cm^2 sec in Iceland for the Postglacial Stage (Bodvarsson, 1955) and $2.5 \cdot 10^{-6}$ cal/cm^2 sec in the Quaternary Kamchatka volcanic zone (Polyak, 1966). Since Iceland is located right on the crest of the active Mid-Atlantic Ridge, it may be reasonable to believe that the volcanic activity has been more active than in island arcs by one or two orders of magnitude. No satisfactory explanation has been given as yet for the discordance of the Kamchatka value with the Japanese value.

Geophysical, geochemical and geological data on active volcanoes are well summarized in the IAVCEI catalogue (Kuno, 1962).

BASALTIC MAGMAS

The petrology of Quaternary volcanic rocks also has a distinct zonality, parallel to the island arc. The variation in chemical composition and mineral assemblage of volcanic rocks would be, in general, subject to the differences in: (*1*) the composition of the primary magma; (*2*) the degree of fractionation such as gravitational, during crystallization; (*3*) the partial pressure of oxygen in magma; and (*4*) the degree of contamination by crustal materials. In spite of variable conditions in (2), (3) and (4), the types of primary basalt magma in Japan have been estimated (e.g., Kuno, 1968). According to the results of these studies, the type of primary magma has been revealed to be persistent in every volcano, at least since the beginning of the Quaternary Period.

The volcanic rocks erupted in San'in district (the Sea of Japan coast area of western Honsyu) and northwestern Kyusyu belong to the "circum-Japan Sea alkaline petrographic province" proposed by Tomita (1935). On the other hand, most of the volcanic rocks in the rest of Japan are thought to have originated from less alkaline basaltic magmas. Tomita (1932) had postulated that there existed two different primitive basaltic magmas. These were called the parent magma of the tracky-basalt—comendite series and the parent magma of the calc—alkaline series, which correspond, respectively, to the alkali-basalt magma and tholeiite magma of Kennedy (1933).

The progress in petrology, in part through studies of basaltic provinces and in part through experimental works, established the reality of Tomita's or Kennedy's two primary magmas (Tilley, 1950), and later investigations have revealed that there are gradual transitions in the rock suites (Green and Poldervaart, 1955). Thus Kuno (1960) proposed a province of high-alumina basalt which is a transitional variety from the alkali-basalt province to the tholeiite province as shown in Fig.40.

As seen in Fig.40, the tholeiite, high-alumina basalt, and alkaline basalt provinces are found to be in a perfect zonality (in Fig.40, for the alkaline basalt magma, Cenozoic rather than Quaternary rocks are shown). Sugimura (1960) attempted to represent the characteristics of the primary magma by an index which is calculated as follows:

$$\theta = SiO_2 - 47 \, (Na_2O + K_2O)/Al_2O_3 \tag{5}$$

where SiO_2 is in weight per cent, and other substances in molecular proportions. Fig.41 is the distribution of the θ values of the area, again clearly showing a zonality. Kuno (1959) attributes the systematic differences in the petrology of the basaltic rocks to the differences in the depth of magma production as will be explained later (see p.182).

The above discussion applies mainly to the east Japan arcs. In the west Japan arcs,

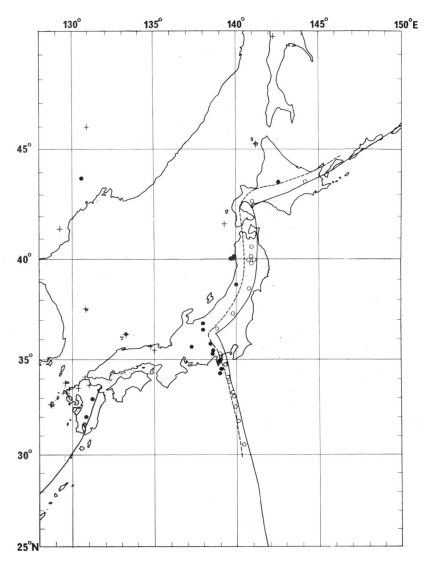

Fig.40. Petrographic provinces of Quaternary volcanoes in Japan and environs. o = tholeiite magma province; ● = high-alumina basalt magma province; + = alkaline basalt magma province. Broken line = boundary of provinces. (After Kuno, 1960.) Solid line = volcanic front.

essential features are similar to the east Japan arcs, but as seen in Fig.40, the volcanoes on the volcanic front are generated from high-alumina basalt rather than tholeiite. In fact, there are considerable differences in the chemical composition of basaltic rocks near the volcanic fronts: the concentrations of Na_2O and K_2O are 1.8−2.1% and 0.2−0.3% in the

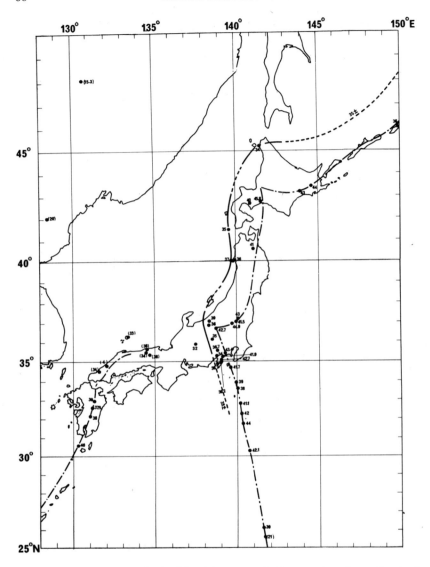

Fig.41. Silica indices (θ-values) of basaltic magmas in Japan and environs. For θ-value, see Sugimura (1968). The dash dot lines indicate the volcanic fronts and the solid lines and dashed lines connect the volcanoes having silica indices of 35 ± 2. Compare with Fig.27. (After Sugimura, 1960.)

east Japan arcs whereas they are 2.4–2.9% and 1.1–1.7% in the west Japan arcs (Sugimura, 1961).

The zonal distribution of petrographic provinces of volcanic rocks can be noticed in other island arcs also. But their details vary from one arc to another (see p.86).

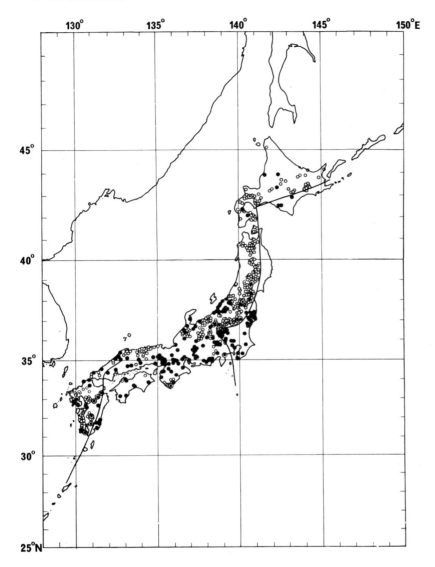

Fig.42. Distribution of hot springs in Japan. (After Fukutomi, 1936.) The lines indicate the volcanic fronts. $\circ = T > 30°C$; $\bullet = T < 30°C$; $\circledcirc$ = group of hot springs with $T > 30°C$.

HOT SPRINGS

A distribution similar to that of the volcanoes was shown by Fukutomi (1936) in the water temperature of hot springs and cold mineral springs as given in Fig.42. It is a striking coincidence that hot springs yielding water warmer than $30°C$ are distributed almost exclusively in the volcanic belt defined in Fig.38. Hot springs must in some way be connected with igneous activity (Horai and Uyeda, 1969). Therefore no high temperature hot springs are expected in non-volcanic regions, and the distribution in Fig.42 supports this view. In the volcanic area west of the volcanic front, almost all hot springs produce high temperature water. It seems to indicate that the underground temperature is generally high in the volcanic belts not only in the immediate vicinities of active volcanoes but also in otherwise ordinary areas. Strong acid springs are also distributed exclusively in the volcanic belts although some are not close to any active volcanoes. Ishikawa (1968) notices that, in Japan, hot springs and geothermal activities are very frequently connected with acidic volcanics of Tertiary age. The authors consider that the age could be limited to the Neogene Period and that this observation of Ishikawa is extremely important, although no definite mechanism has been proposed to account for it.

In Wakayama prefecture in the Kii Peninsula (Fig.63, p.92), there are many hot springs yielding the water $33-90°C$ in temperature. Since the area is out of the volcanic belts, this fact may be regarded as anomalous. One of the heat flow measurements in the same area indicated $2.12.10^{-6}$ cal/cm^2 sec, and geologically the Wakayama area belongs to the "Outer Zone" of southwest Japan where the crust is dotted with Neogene acidic igneous rocks. The high temperature springs in Wakayama may be related to the activity of this type. This particular area has been known for its swarm of earthquakes (e.g., Miyamura et al., 1966). In this respect, it may be worth noting that the Nankai Trough area shows abnormally high heat flow (Fig.43).

HEAT FLOW

Terrestrial heat flow Q is defined by:

$$Q = \kappa \frac{dT}{dz} \tag{6}$$

where κ and $\frac{dT}{dz}$, are the thermal conductivity and the geothermal gradient in the crust.

Q is usually given in the units 10^{-6} cal/cm^2 sec which is often abbreviated as H.F.U. (heat flow unit) (Simmons, 1966). At the 95% confidence level, the world's mean heat flow is 1.5 ± 10% H.F.U. and the average over the continents does not differ significantly from that over the oceans. Heat flow values are definitely correlated with major geological features: on land, heat flow is uniform and slightly sub-average in Precambrian shields (Q = 0.92 ± 0.17 H.F.U.) and other stable areas whereas it is higher in the Mesozoic—

Cenozoic orogenic areas (Q = 1.92 ± 0.49 H.F.U.); at sea, heat flow is uniform and sub-normal in oceanic basins (Q = 1.28 ± 0.53 H.F.U.) and higher over mid-oceanic ridges (Q = 1.82 ± 1.56 H.F.U.). It is generally low in oceanic trenches (Q = 0.99 ± 0.61 H.F.U.). Details of these features are found in the reviews by Lee and Uyeda (1965) and Simmons and Horai (1968).

Island arc areas are quite anomalous in heat flow. Heat flow measurements in the area of Japan and environs have been carried out since 1957 by the Earthquake Research Institute of the University of Tokyo, the Japan Meteorological Agency, and the Hydrographic Department (Uyeda et al., 1958; Horai, 1959, 1963a, b, c, 1964; Uyeda and Horai, 1960, 1963a, b, 1964; Uyeda et al., 1961, 1965; Horai and Uyeda, 1963; Yasui et al., 1963). In 1966, the Scripps Institution of Oceanography, University of California, joined the cooperation by sending the R.V. "Argo" into the western Pacific (Vacquier et al., 1967a). The measurements in the Sea of Japan and the Sea of Okhotsk have been conducted by M. Yasui and collaborators (Yasui and Watanabe, 1965; Yasui et al., 1967a, b; 1968a, b). Cooperation of the Japanese group and Lamont (now Lamont-Doherty) Geological Observatory, Columbia University, in heat flow measurements has also been active in the last few years in the western Pacific, in particular in the areas of the Ryukyu Arc (Yasui et al., 1970), and the Philippine Sea (Watanabe et al., 1970). Mizutani et al. (1970) made a series of land measurements in South Korea through a cooperative project with the Korean Geological Survey. Owing to these endeavours, as far as heat flow is concerned, the Japanese area is now one of the most thoroughly studied areas of the world. These data are summarized in Fig.43.

Notable features of heat flow distribution in the Japanese Islands are the existence of a low heat flow zone (average Q = 0.70 ± 0.15 s.d., H.F.U.) in the Pacific Ocean side of the east Japan island arcs and that of a high heat flow zone in the Sea of Japan side and the Fossa Magna region (average Q = 2.01 ± 0.38 s.d., H.F.U.). This distribution coincides almost exactly with the distributions of volcanoes and hot springs illustrated in Fig.38 and 42 and consequently with the general trends of the youngest orogeny (Fig.75, p.114). Heat flow stations are usually taken deliberately away from the immediate vicinity of active volcanoes or geothermal areas. Yet, the heat flow in the Sea of Japan side and the Fossa Magna region is generally high. The contrast of heat flow on the land of Japan is extended to the oceanic area. The heat flow in the northwest Pacific Ocean Basin off the northeast Japan and Izu–Bonin arcs is uniform and sub-normal (average Q = 1.15 ± 0.37 H.F.U.). There is a slight but definite decrease of heat flow in the trench areas. It was also found that the major topographic highs such as the Shatsky Rise (northwest Pacific) and Emperor seamounts also have sub-normal heat flow values. These results show that geothermally the northwestern Pacific Basin is in sharp contrast with the eastern Pacific where the heat flow is much less uniform and high in average (average Q = 2.2 H.F.U., Vacquier et al., 1967b). On the other hand, the high heat flow in the Sea of Japan side is extended to the whole of the Sea of Japan and the southern part of the Sea of Okhotsk. As shown in Fig.38 no active volcanoes exist in these marginal seas at present

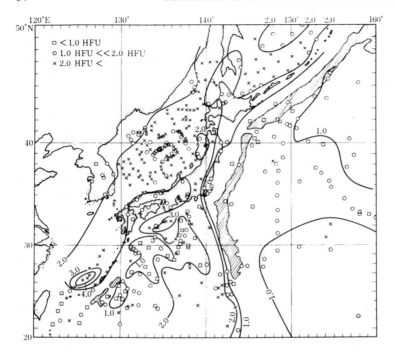

Fig.43. Distribution of heat flow in and around Japan. (After Uyeda and Vacquier, 1968; Yasui et al., 1970a; Watanabe et al., 1970; Mizutani et al., 1970.)

and the crust of the deep basins are oceanic (see p.24). From such data alone, no high heat flow would be anticipated in these seas. Therefore, the observed high heat flow values are anomalous and require explanation. In Chapter 3 (p.176) a model of marginal sea basins as a spreading plate will be presented to explain these features. To summarize, it may be postulated that, in an active island arc area, heat flow is low on the ocean side and high on the continent side. Low heat flow on the ocean side may be explained by the descending mantle flow. Various chemical or metamorphic processes accompanying the downflow are considered as endothermic, so that they would play a role of negative heat source, entailing the local enhancement of low heat flow zone (Uyeda and Horai, 1964). However, the high heat flow inside the arc is difficult to understand (see p.168). From this point of view, a thorough survey of the Sikoku–Philippine Basin would be of particular interest, because it is a sea behind an active arc of Izu–Marianas and at the same time a sea in front of the arcs of west Japan and the Philippines. The data obtained so far in these areas (Fig.43) indeed indicate a complex distribution of heat flow (Yasui et al., 1970; Watanabe et al., 1970). Contrast of low and high heat flows on both sides of a trench–arc system can be fairly clearly observed for the Ryukyu Islands and for the Izu–Mariana Islands. In particular, the very high heat flow in the Okinawa Trough behind the Ryukyu Arc is remarkable. This trough may be regarded as an embryo of a marginal sea basin in the model presented on p.176. On the other hand, the Philippine Sea Basin and Sikoku Basin (Fig.8) seem to have numerous adjacent areas of high and low heat flow. Their distribution does not necessarily coincide with such topographic features as the Kyusyu–Palau Ridge. One anomalous feature to be noted is the high heat flow in the area of Nankai Trough, south of Sikoku. High heat flow in such a topographic depression in front of an arc is not known elsewhere. This observation again shows that southwest Japan is not a typical island arc. As noted earlier, the troughs west of the Izu–Mariana Arc may be another example of extensional spreading (embryo of a marginal sea). But a few heat flow measurements in the inner Mariana Trough (Watanabe et al., 1970) gave low values. The information that may be obtained by a more detailed heat flow survey will be of key importance in clarifying the origin and development of the Philippine Sea.

Distributions of heat flow and seismicity appear to be correlated in a complex manner. If we compare Fig.25 with Fig.43, it can be seen clearly that the activity of shallow earthquakes and heat flow are negatively correlated: the highest seismicity in the sea off northeast Japan is in the area of the lowest heat flow and vice versa. This negative correlation is important since it is in conflict with the idea that earthquakes are caused by excess supply of heat flow from the earth's interior (e.g., Matuzawa 1953). But in the case of deep earthquakes, the seismicity and heat flow appear to be correlated positively in a larger scale, i.e., the zones of high heat flow in Fig.43 roughly coincide with the zone bounded by the 155-km isobath and the broken curve in Fig.27. If we look at the situation more closely, however, there are indications that the correlation may again be negative. Yasui et al. (1968b) report such a local negative correlation in the Sea of Okhotsk. Such a relation, if true, may be of considerable significance in interpreting the processes occurring under the island arcs. To establish such a complicated correlation, more accurate delineation of both deep seismic foci and heat flow would be required than is presently available.

The heat flow at a depth d, may be expressed, in a steady state, as:

$$Q\,(d) = Q_{surface} - \int_{0}^{d} A(Z)\,dz \tag{7}$$

where $A(Z)$ is the rate of heat generation per unit time and volume. When d is equal to the depth of the M-discontinuity, the heat flow may be called the Moho-heat flow. The A-value in the crust is difficult to estimate even if we consider only the radiogenic heat. Here, A-values of the granitic layer and basaltic layer are assumed as $4 \cdot 10^{-13}$ cal/cm^3 sec and $1 \cdot 10^{-13}$ cal/cm^3 sec, respectively, and the crust is assumed to consist of one part of granitic layer and one part of basaltic layer on land, but to be entirely basaltic in oceanic areas. The Moho-heat flow of the area, thus estimated, is shown in Fig.44. In the Sea of Japan area, such a high heat flow over 2 H.F.U. is expected to come from the mantle while very little heat should come from the mantle in the low heat flow zone. It is even negative in the eastern zone. It must be noted, however, that, even if the radioactive heat in the earth is taken correctly, the Moho-heat flow thus computed would not represent

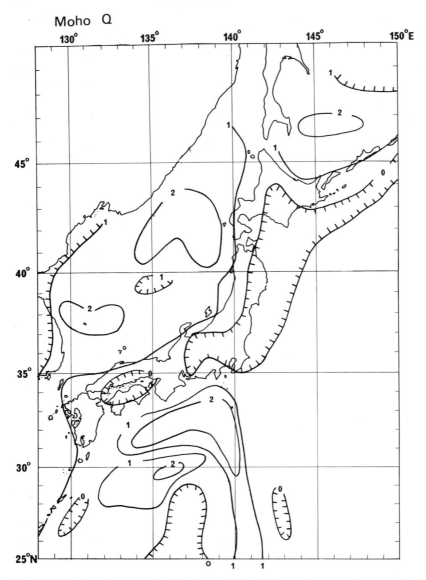

Fig.44. Moho-heat flow in 10^{-6} cal/cm^2 sec. (After T. Watanabe, personal communication, 1968.)

the actual heat flowing through the M-discontinuity when some exothermic or endo-thermic processes other than radioactive decay take place within the crust. Meta-morphism may be an example of endothermic reaction, playing the role of a thermal sink. In order to avoid the negative heat flow and the decrease of temperature at depths, it may be that we need to take into account some regional metamorphism which absorbs heat (see p.162).

As will be explained later (see p.157), it is possible to obtain some idea about the temperature distribution in the crust and upper mantle from the surface heat flow data. The data in and around the Japanese area indicate that isotherms are raised under the inner side of the arc where the heat flow is high and depressed under the outer zone of low heat flow. It may be worth noticing that the temperature in the upper mantle is higher under continent side than under ocean side in the island arc type continental margin. It is in contrast to the otherwise distribution of upper mantle temperature deduced from the premise of equality of heat flow and inequality of crustal structure of continents and oceans (e.g., MacDonald, 1963; Clark and Ringwood 1964; Lubimova and Magnitzky, 1964; Pollack, 1967). The cases discussed by these authors are probably valid for ordinary continental margins but not for active island arc type margins. This difference in the distribution of upper mantle temperature seems to have a profound significance as will be shown in Chapter 3.

GEOMAGNETIC FIELD

Fig.45 is the map showing the total geomagnetic field intensity of the area around Japan. The areas east of the northeastern Honsyu, southwestern Honsyu, the Sea of Japan and the Sea of Okhotsk have been mapped by Uyeda et al. (1964, 1967), Matsuzaki (1966), Tomoda and Segawa (1967), Segawa et al. (1967), Tomoda et al. (1967), Yasui et al. (1967c, d), Segawa (1968) and Isezaki and Uyeda (1970). The land magnetic surveys have been carried out by the Geographical Survey Institute of Japan (Fujita, 1966; Tazima, 1966b). Here, however, only marine data are presented. Some data from the Scripps Institution of Oceanography (R. Warren, personal communication, 1962), the Lamont-Doherty Geological Observatory (J. Heirtzler, personal communication, 1966) and the U.S. Naval Oceanographic Office (Bracey, 1963, 1966) were incorporated in this compilation. It may be observed clearly that the regional trend in the area is represented by southwest–northeast contours, and, after reducing the regional trend, lineations with several hundred gammas exist in the Pacific east of northeast Japan as shown in Fig.46. These lineations are similar to those found in the eastern Pacific by V. Vacquier and his colleagues (Mason, 1958; Vacquier et al., 1961; Mason and Raff, 1961; Raff and Mason, 1961). The lineations are not apparent in the sea west of the Izu–Mariana Arc; but vaguely noticeable in the Sea of Japan and the Sea of Okhotsk. The amplitude of the anomaly in these marginal seas is much smaller than that in the Pacific Ocean Basin. On the land linear anomalies do not seem to exist. It may be said that the lineations die out as land is approached. This is in agreement with the observations in the eastern Pacific and elsewhere.

The cause of such magnetic lineations must be of great importance. The tape-recorder model of a spreading ocean floor by Vine and Matthews (1963) appears to be successful in explaining the origin of lineations in many areas. If the Vine–Matthews hypothesis is to be applied to the lineations found in the northwestern Pacific, one has to accept that

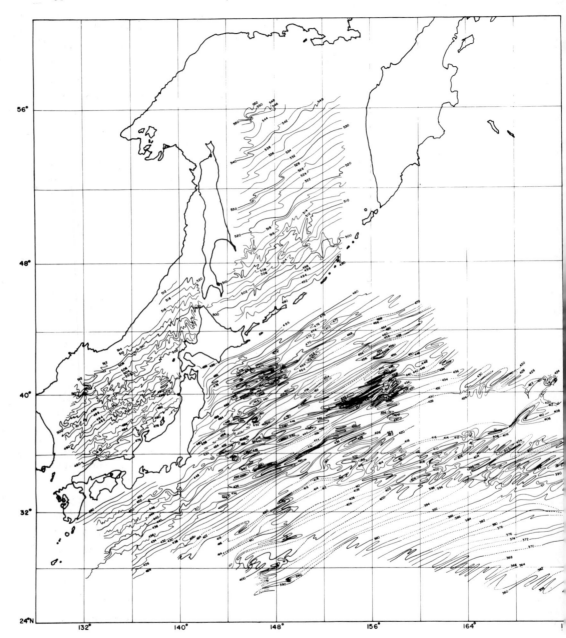

Fig.45. Total geomagnetic field in and around Japan in units of 100 gammas. (After Yasui et al., 1967c, d; Uyeda and Vacquier, 1968.)

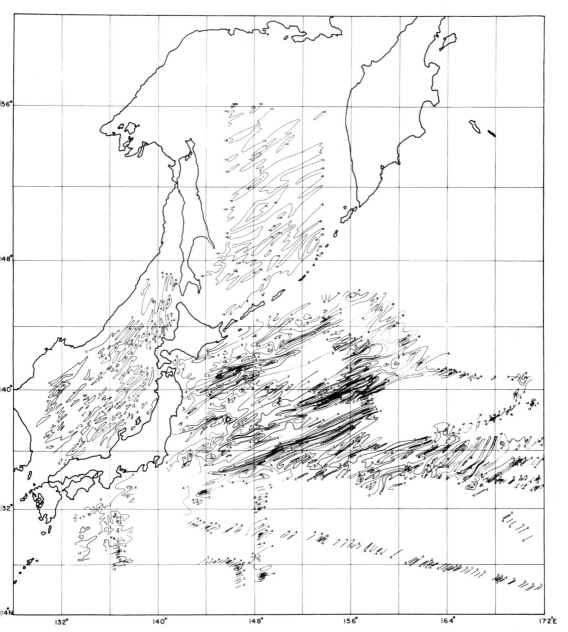

Fig.46. Anomaly of total geomagnetic field around Japan in unit of 100 gammas. (After Yasui et al., 1967c, d; Uyeda and Vacquier, 1968.)

they are very far indeed from the present day sources of the ocean floor, i.e., the East Pacific Rise and the Pacific—Antarctic Ridge. The Shatsky Rise and the Emperor seamount ridges are considered inactive in the recent geological past because of the observed subnormal heat flow (see p.63). In this connection, however, attention must be paid to the actual manner in which the lineations in the western Pacific are linked or not linked with those in the eastern Pacific, in particular with the lineations in the northeast Pacific (Peter, 1966; Elvers et al., 1967; Uyeda et al., 1967; Hayes and Heirtzler, 1968; Pitman and Hayes, 1968; Erickson and Grim, 1969; Grim and Erickson, 1969). If the southwest—northeast trending northwest Pacific lineations were produced on the East Pacific Rise and have travelled over the entire Pacific Ocean Basin, they should simply be the oldest part of the Pacific Basin. These lineations should not be linked with those in the northeast Pacific, because the latter are considered to have their source-ridge in the north, now under the Bering Sea Basin or Alaska (Pitman and Hayes, 1968). If, on the other hand, the lineations in the northwest Pacific and the northeast Pacific are linked across the Emperor seamounts, one would have to accept that the source of the lineations we observe in the area may now be in Siberia. Moreover in such a case, the direction of the ancient spreading and that of the present movement of the Pacific Ocean floor are 180° out of phase: a situation now supposed for the Aleutian Arc. So far it has not been possible to make a linkage of the lineations; the magnetic profiles in the northwest Pacific have not been correlatable with those in the northeast Pacific. The distribution of the opaque layer (Ewing et al., 1968) indicates that the northwestern basin is older in age than the northeastern basin, so that the lineations should not be linked (Hayes and Heirtzler, 1968). As will be described in the section on paleomagnetism (p.75), paleomagnetic studies on seamounts have indicated that the northwest Pacific Basin has moved northward by 20–30° in latitude since Cretaceous times.

Recently, Hayes and Pitman (1970) have discussed the magnetic anomalies in the north Pacific in a comprehensive paper. According to these authors, the tectonic evolution of the north Pacific was dominated by migration of ridge—ridge—ridge type triple junctions (McKenzie and Morgan, 1969). The nature of the lineations off northeast Japan is still not completely clear.

ELECTRICAL CONDUCTIVITY

With regard to the electromagnetic aspects of the island arc structure, the pronounced anomaly in the electrical conductivity in the mantle beneath Japan, as deduced from the anomalous time variations in the geomagnetic field (e.g., Rikitake, 1959a, 1966), must be mentioned. When the earth's magnetic field is changed by some extraterrestrial causes such as the solar daily variation, magnetic storms and other disturbances, some electric currents are induced within the earth, giving rise to the secondary magnetic fields. The magnitude and the phase of the secondary fields are determined by the electromagnetic properties of the earth's interior, in particular the distribution of electrical conductivity.

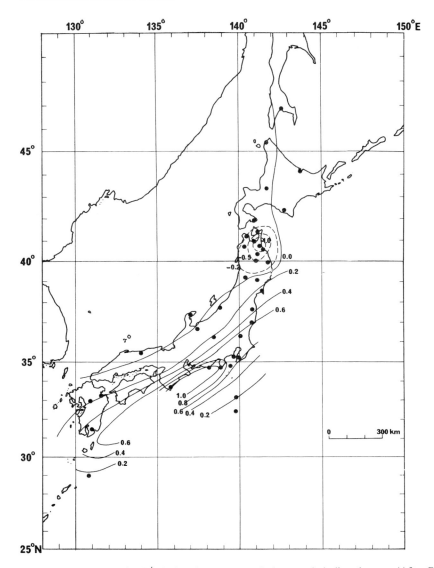

Fig.47. Distribution of $\Delta Z/\Delta H$ for the geomagnetic bays and similar changes. (After Rikitake, 1966; Kato, 1968.)

It has been a classic practice to estimate the electrical conductivity of the mantle from the analysis of the time variations of the geomagnetic field (e.g., Tozer, 1959; Rikitake, 1966). According to these investigations, the electrical conductivity in the mantle increases rather sharply at the depth of about 400 km (the C-layer) from 10^{-15} e.m.u. to 10^{-12} e.m.u. In addition to the effect of general rise of temperature, the olivine–spinel transformation may be the cause of this increase of electrical conductivity (Rikitake, 1959b; Akimoto and Fujisawa, 1965).

According to T. Rikitake and his colleagues (Rikitake et al., 1952, 1953, 1956, 1958, 1959, 1962; Rikitake, 1956, 1959a, b, c, 1966), the modes of the geomagnetic variations are anomalous in Japan. The anomaly can be best seen in the case of the geomagnetic bay-type variation with the period of one hour or so. Fig.47 shows the distribution of $\Delta Z/\Delta H$ for bays and similar changes, where ΔZ (positive downward) and ΔH are the vertical and horizontal components of the changing field. It is clearly observed that $\Delta Z/\Delta H$ is anomalously large in the southern marginal area of central Japan. Such a large $\Delta Z/\Delta H$ cannot be expected for a horizontally layered earth: for the type of external variation, $\Delta Z/\Delta H$ must be small and negative. The observed anomaly in $\Delta Z/\Delta H$ should be caused by an anomaly in the electrical conductivity in the earth. Anomaly in electrical conductivity thus deduced is often called the conductivity anomaly or simply C.A. Intensive observations during the Upper Mantle Project (U.M.P.) and International Quiet Sun Years (I.Q.S.Y.) revealed that an equally intense anomaly with the opposite sign exists in the northern end of Honsyu (Kato, 1968). The above two anomalies are called the Central Japan Anomaly and the Northeastern Japan Anomaly, respectively. In the Sea of Japan side of Honsyu, $\Delta Z/\Delta H$ is small. Electromagnetic theory predicts that the Z-component due to the electromagnetic induction within the earth will tend to compensate the Z-component of the externally applied variations. The smallness of ΔZ in the Sea of Japan side accords the idea that the compensation is effective in this area. This would correspond to a high electrical conductivity and consequently to a high subterranean temperature underneath the Sea of Japan side of Honsyu, a result expected from the heat flow distribution (see p.64). The Central Japan Anomaly, however, indicates a high degree of overcompensation which is very difficult to explain. On the other hand, the modes of the variation of longer period, such as Sq (solar daily variation) and Dst (magnetic storm) indicate that the electrical conductivity beneath central Japan is anomalously low at a greater depth of, say, 400—700 km. On account of these observations Rikitake (1956) proposed a rather complicated configuration of electrical conductivity in the upper mantle where a complex loop of highly conducting channels comes up from a great depth (400 km) to a shallow depth (200 km) under the anomaly, whereas a low conducting material wedges into the high conducting layer to a depth of about 700 km. The wedge of the low conducting materials might be related to the low attenuating zone of Utsu (1966, 1967) and Utsu and Okada (1968) mentioned earlier (p.32) and to the downgoing convection currents as Rikitake pointed out. In order to clarify the distribution of subterranean electrical conductivity of the area, it appears that $\Delta Z/\Delta H$ in the oceanic areas must be studied by sea-bottom magnetic measurements.

Some authors (e.g., Roden, 1963) pointed out that the anomaly may be explained by the effect of the sea water of which electrical conductivity ($\sim 10^{-11}$ e.m.u.) has orders of magnitude higher than that of rocks. But it has been further noticed that the effect of the highly conducting layer in the mantle should also be important in producing the anomaly in the short period magnetic variations. Other things being equal, the depth of the mantle conducting layer is determined by the temperature of the mantle. Therefore, it can be expected that the electrical conductivity anomaly reflects the temperature distribution in the upper mantle. An attempt to explain the observed anomalies in terms of the upper mantle temperature will be made in Chapter 3 (see p.165).

PALEOMAGNETISM

Paleomagnetism is the study of the ancient geomagnetic field by means of the natural remanent magnetization of rocks and other objects (e.g., Irving, 1964). Investigation of magnetic properties of rocks has been one of the most active branches of geophysics in Japan. As early as in the 1920's, Matuyama (1929) discovered a number of occurrences of naturally reversely magnetized volcanic rocks of Cenozoic age, and proposed the then pioneering idea that the earth's magnetic field has undergone reversals in geologic history. In the 1940's Nagata (1961) made a series of classic investigations on rock magnetism and contributed, together with other pioneers such as Thellier (1951), Graham (1949), and Neel (1949), in providing a physical basis on the possible use of natural remanent magnetization of rocks as fossils of the ancient geomagnetic field. Basic physical investigations as well as some paleomagnetic application of rock magnetism in Japan has been actively continued by T. Nagata and his colleagues and N. Kawai and his colleagues. Self-reversal of thermoremanent magnetization was discovered on the Haruna dacite (Nagata et al., 1951) and its physical mechanism was investigated in detail (Uyeda, 1958; Ishikawa and Syono, 1963). As is now well known, the attempt of explaining all the natural reverse magnetization of rocks in terms of self-reversal rather than by the reversals of earth's field was unsuccessful. Thus, a detailed paleomagnetic study was made to determine the period of the last reversal of the earth's field by the volcanic rocks of the Izu—Hakone region (Nagata et al., 1957) and confirmed Matuyama's early proposal that it was in the Early Quaternary or the latest Tertiary (Matuyama, 1929). Polarity history of the earth's field has been continued by Momose (1963) and Nomura (1967).

Due to the violent tectonism, however, Japanese rocks much older than, say, Pliocene were considered unsuitable for paleomagnetic studies, so that relatively little attention was paid to them until Kawai et al. (1961, 1969) noticed the striking possibility of bending of the Japanese Islands in the Cretaceous time: first they pointed out, on the basis of paleomagnetism on 28 sets of rocks collected from different parts of Japan, that the paleomagnetic directions of rocks of pre-Tertiary age from southwest Japan differ systematically from those from northeast Japan, whereas all the paleomagnetic directions are more or less coincident for post-Paleogene rocks over the whole Japanese Islands (Fig.48). They suggested that the observed systematic discrepancy in the paleomagnetic directions was due to the major bending of the Japanese Islands in Late Mesozoic or Early Tertiary times. Later, with the aid of K—A dating, the manner of the proposed bending was examined in detail. As shown in Fig.48, the paleomagnetic directions of rocks from

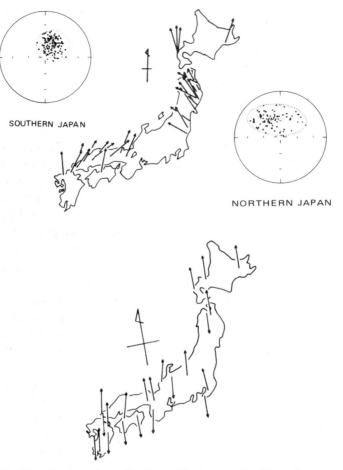

SOUTHERN JAPAN

NORTHERN JAPAN

Fig.48. Paleomagnetic declinations of Japanese rocks. Upper figure for rocks of pre-Tertiary age and lower figure for Tertiary rocks. (After Kawai et al., 1961, 1969.)

northeast Japan are characteristically distributed in an ellipse, while those from southwest Japan possess a circular distribution. K–A age information indicated that the ages of the rocks in the ellipse systematically increase toward the westerly part of the ellipse from 80 m.y. to 120 m.y. From this, it was inferred that the anti-clockwise bending of northeast Japan relative to southwest Japan took place during the period from 120 m.y. to 80 m.y. with the mean rate of rotation of $2.9°$/m.y. This possibility of major bending may be of great importance in the tectonic development of the Japanese Islands. Kawai and Nakajima (1970) discuss this possibility from the viewpoint of the length shortening of the arc.

Recently, Sasajima et al. (1968) made a study on Paleogene volcanics from southwest Japan. The paleomagnetic pole positions deduced from these rocks are found in southern

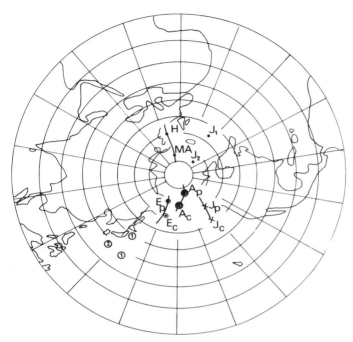

Fig.49. Virtual pole positions. Jp = Paleogene rocks of southwest Japan; Jc = Cretaceous rocks of southwest Japan; Ep = Paleogene rocks of Eurasia; Ec = Cretaceous rocks of Eurasia; Ap = Paleogene rocks of North America; Ac = Cretaceous rocks of North America; J_1, J_2 = seamounts off Japan; H = Hawaiian seamounts; MA = Midway atoll. (After Irving, 1964; Vacquier and Uyeda, 1967; Sasajima et al., 1968; Francheteau et al., 1970.)

Alaska as shown in Fig.49. The position is at a reasonable place on the polar path previously deduced from Cretaceous red shales from southwest Japan (Nagata et al., 1959). Comparing the poles from Japan and from Eurasia, Sasajima et al. (1968) postulated that the Japanese Islands have drifted north or northeastward by about 1,000 km or so and rotated clockwise by about 20° since the Cretaceous, relative to the Eurasian continent. It appears to be a reasonable inference from Fig.49, although the argument rests on the appropriateness of using the Eurasian poles which have been obtained largely from European data. Alternatively, it may be more reasonable to say, until more reliable pole positions become available from the nearby continental area, that the polar path from southwest Japan represents at least partially its clockwise rotation.

Another line of paleomagnetic study is that of seamounts. The direction and intensity of magnetization of a uniformly magnetized seamount can be computed if its shape and associated magnetic anomaly are known. This technique has been applied on a number of seamounts near the Japanese Islands (Uyeda and Richards, 1966; Vacquier and Uyeda, 1967; Yasui et al., 1970). The location of the seamounts and the paleomagnetic pole positions deduced from them are shown in Fig.49. The poles from seamounts in the

Pacific Basin are found in the north Atlantic. Radiometric and paleontological evidence indicated that these seamounts originated probably in Cretaceous times. Although dependent again on the method of choosing the reference Cretaceous pole for the Asiatic continent, the results described above suggested that the Pacific Basin has migrated northward by a few tens of degrees in latitude. This result seems to harmonize with the movement of the Pacific Ocean floor presumed in the sea-floor spreading hypothesis. A group of three seamounts of unknown age in the Philippine Sea (Sikoku Basin) gave pole positions closer to the present spin axis than the seamounts in the Pacific Basin. Recently, Francheteau et al. (1970) reviewed the results of paleomagnetic studies of 49 seamounts in the eastern Pacific. They proposed a preliminary polar path curve relative to the northeastern Pacific as shown in Fig.49. Again some 30° northward movement may be suggested for the northeastern Pacific Basin since the Cretaceous. Comparing this polar curve and the average pole for the Japanese (Cretaceous) seamounts, it may be suspected that the plates of the northwestern and northeastern Pacific have not always been rigidly connected.

ZONAL ARRANGEMENTS IN OTHER ISLAND ARCS

As emphasized on p.1, it is not only for the Japanese Islands but also for other island arcs that various geological and geophysical features are arranged in zones with a definite order: approaching the continent from the ocean basin, the order is, trench, negative gravity anomaly, positive gravity anomaly, axis of islands, volcanic belt, and deep earthquake zone. As will be treated in Chapter 3, the authors believe that this regularity originates from a common cause, that is, the descending mantle flow in the island arc areas. The model of a descending mobile lithosphere as developed by Isacks et al. (1968) is highly favoured by the present authors. In this section, a brief review on the regularity mentioned above will be made about other island arcs. Recent series of works on global seismicity and earthquake mechanism by the scientists of Lamont-Doherty Geological Observatory (Sykes, 1966; Isacks et al., 1968, 1969; Barazangi and Dorman, 1969; Isacks and Molnar, 1969; Katsumata and Sykes, 1969; Molnar and Sykes, 1969; Fitch and Molnar, 1970) has greatly contributed in elucidating the regularity of zonal structures for various island arcs.

The position of trenches relative to the Pacific Ocean Basin is reversed in some arcs, such as the Indonesia, Melanesia, West Indies, and Scotia arcs. Apparently the system of the flow in the mantle or of the mobile lithosphere blocks is not simple. In fact, all these arcs are situated at the portion where two oceans meet. The New Hebrides Arc terminates at its southern end where the arc bends rather sharply toward the northeast. On its extension beyond the Fiji Islands, the Tonga—Kermadec Arc starts at a similar bending. These two arcs have the trenches on reversed sides. This situation, together with the similar ones in the West Indies and Scotia arcs, can be explained by the arc-to-arc

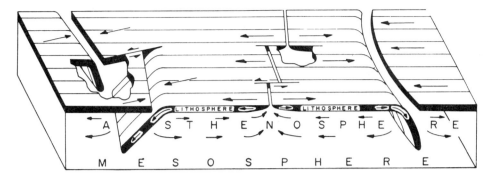

Fig.50. Block diagram illustrating the configurations of lithosphere, asthenosphere and mesosphere in the new global tectonics. The arc-to-arc transform fault on the left is supposed to be the case for the Tonga-Fiji-New Hebrides area. (After Isacks et al., 1968.)

transform faults as is neatly illustrated in Isacks et al., (1968, fig.1). Their figure is reproduced in Fig.50.

It happens that both of the two classical island arcs that have been investigated most intensively by western scientists are "reversed" arcs: one is the Indonesian Arc (East Indies) studied by the Dutch, and the other is the West Indies Arc studied mainly by the Americans. As an example, Fig.51 shows the distribution of gravity anomaly and the Late Cenozoic fold zones in the Indonesian arc area. Volcanoes and deep and intermediate earthquake epicenters are also shown. It can be seen clearly that the zonal pattern holds true in this area. Recently Santo (1969a, b, 1970), using the *Preliminary Determination of Epicenters* and/or the *Earthquakes Data Report* by the U.S. Coast and Geodetic Survey for 1964–1969, has made a series of studies on the global seismicity. He noticed (Santo, 1969b) that the distributions of foci in planes perpendicular to the arc at various portions between Java and Burma can be expressed by a common curve. It was maintained that the seismicity under the Andaman Islands and Burma can be considered as of island arc type, although the deepest foci in these regions is shallow (~ 300 km) as compared with other parts having deep oceanic trenches. Fitch and Molnar (1970) have made extensive determinations of focal mechanisms of intermediate and deep shocks in the Indonesia–Philippine region. They confirmed that at intermediate depth the axes of minimum compressive stress are aligned parallel to the seismic plane (down-dip extension) whereas at greater depths maximum compressive axes are parallel to the seismic plane (down-dip compression, see Isacks and Molnar, 1969). In the regions where the seismic zones are not planar, such as the Celebes and Banda seas, a striking reorientation of stress axes was found. These authors postulate that these results are in agreement with the basic ideas of plate tectonics.

As for the heat flow, systematic investigations in the Indonesian region has been made only in the Indian Ocean west of Sumatra (Vacquier and Taylor, 1966), where the heat flow is subnormal in the Sumatra Trench area and above normal off the trench. There are

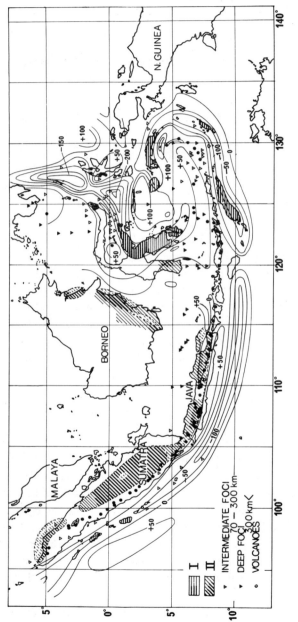

Fig.51. Gravity anomaly (isostatic anomaly in milligals) and Late Cenozoic fold zones in Indonesia. (Arranged from Umbgrove, 1947; and Vening Meinesz, 1964.) *I* = Tertiary fold zone; *II* = Tertiary sedimentary basin with gentle folding.

only a few heat flow data in the marginal seas such as the Sulu and Celebes seas (Nagasaka and Kishii, 1970). These preliminary measurements are definitely too scarce to draw any conclusion about heat flow distribution in these seas. Measurements in the land area of the arc as well as in the marginal seas, such as the Banda Sea and Sunda Sea, are highly desirable.

The Miocene fold zone indicated in Fig.51 can be divided into two belts: the outer fold belt, showing the Alpine type thrusts; and the inner fold belt, where the folding is gentler and lasted until the post-Pliocene period. The inner belt is of the type called idiogeosyncline and, in having thick Miocene strata and oil fields, highly resembles the Uetu geosyncline in northeast Japan (see p.101) in form, degree and age of folding. Quaternary volcanoes are most densely distributed on the outer rim of the inner belt, again showing a similarity with northeast Japan. One may also notice that the belt of negative gravity anomaly coincides with the outer Alpine type fold belt. In the case of northeast Japan, the axis of the negative gravity anomaly is just inside the Japan Trench (Fig.14). If the situations are similar to the Indonesian Arc, it can be expected that active Alpine type thrusting occurred in Miocene time and is even in progress under the zone of negative gravity anomaly in the area between the coast of northeast Japan and the Japan Trench. The radiation pattern of the first motions of earthquakes occurring in this area supports this inference (see p.198). Similar underthrusting of the seaward block beneath the landward block has been noticed in other arcs, such as the Tonga—Kermadec Arc (Isacks et al., 1969), and the Aleutian Arc (Stauder, 1968).

It is well known that similar negative gravity anomaly belts exist in other island arcs, too (see p.21). These anomalies say that isostasy is not realized in these areas. It seems that some light materials are being forced down in these places by some forces acting against buoyant force. Vening Meinesz's down-buckling hypothesis, or Kuenen—Umbgrove's tectogene hypothesis are based on the observation of the negative gravity anomaly associated with the trenches. These authors as well as Griggs (1939) postulate that the forces causing or supporting the anomaly is the convection current in the mantle. There is no doubt about the existence of gravity anomalies, but as to the reality of the bulging crustal structure postulated by the above hypothesis there seems to be considerable controversy. For this problem, crustal studies by seismic methods must be incorporated. Fig.52 is an example of such an attempt for the Puerto Rican "Arc" (Talwani et al., 1959). The crust under the trench is thickened in the figure but is distinctly thinned on both sides of the trench, suggesting the existence of normal faults. A similar tendency was found for the Tonga Arc (Talwani et al., 1961). In the cases of the Mindanao Trench (Worzel and Shurbet, 1955) and the Cayman Trench (Ewing and Heezen, 1955) even the thickening of the crust under the trench was not required to explain the observations. Based mainly on these observations, some scientists of the Lamont-Doherty Geological Observatory postulate that the forces acting in the crust under trenches is tensional, and regard the hypothesis of tectogene as "beautiful but unsupportable" (Heezen, 1967). The observation of normal faults by seismic profiling (Ludwig et al., 1966; Fig.95) in the Japan Trench and the studies of earthquake mechanism (Isacks et al., 1968) are in favour of the tension hypothesis. As will be mentioned in Chapter 3, however, the tensional features observed are interpreted as being due to the bending of a sinking slab of lithosphere by other scientists of the same observatory (Isacks et al., 1968).

Under the Kurile Trench, where U.S.S.R. scientists have made detailed investigations (see p.26), crustal structures as illustrated in Fig.19 have been presented. Pronounced thickening of the crust just inside the trench may be observed.

The hypothesis of underthrusting of the lithosphere with high Q and high V under island arcs was postulated for the east Japan arcs (Utsu, 1967; Utsu and Okada, 1968; Fig.24) and for the Tonga Arc (Oliver and Isacks, 1967; Fig.53). Their results were obtained by the analysis of transmission of seismic waves. Such a model appears to be consistent with the earlier models based on explosion seismology. For instance, the shape of the 6.6–7.1 km/sec layer drawn in the right half of Fig.54, which is included in the results from the Puerto Rico Trench by Officer et al. (1959), seems to represent this

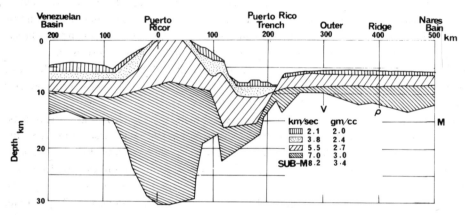

Fig.52. Crustal cross-section of the Puerto Rican Arc. (After Talwani et al., 1959.) V = velocity of P-wave; ρ = density.

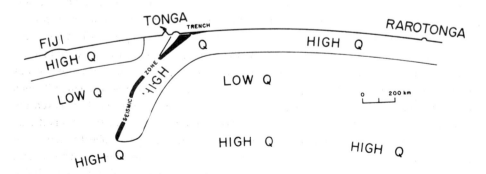

Fig.53. Hypothetical section through Fiji–Tonga and Rarotonga. (After Oliver and Isacks, 1967.)

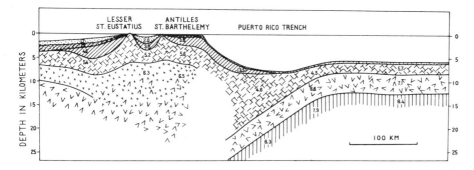

Fig.54. Structure based on explosion seismology across the Puerto Rico Trench. (After Officer et al., 1959.)

consistency. The same appears to hold with Fig.55 for the Peru—Chile Trench (Fisher and Raitt, 1962) and Fig.56 for the trench off southeast Alaska (Shor, 1966). It may be added that these models were prepared before the advent of the sinking lithosphere model.

Let us now examine the zonality of the deep earthquakes. The angles of dip of the seismic plane under various arcs are listed in Table IV. Recently, extensive redetermination of the focal positions of deep earthquakes has been made (Sykes, 1966; Katsumata

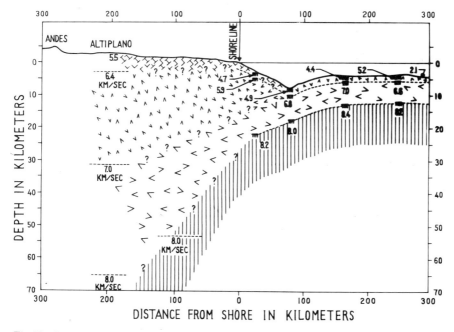

Fig.55. Structure based on explosion seismology across the Peru—Chile Trench. The patterns are adapted to Fig.54. The sense of right and left in the section is changed also for adapting. (After Fisher and Raitt, 1962.)

TABLE IV

Characteristics of island arcs

Region	Maximum depth of trench[1] (m)	Distance from axis of trench to front of volcanic belt[2] (km)	Depth of seismic plane under front of volcanic belt[3] (km)	Dip of seismic plane for intermediate foci[4]
New Zealand				
Tonga, Kermadec	10,882		100	58,64
New Hebrides,				
Solomon	9,165			42
New Britain,				
New Guinea	8,320			
Indonesia	7,450	300		35
Celebes, Halmahera				
Sangihe				
Philippines	10,497	180–200		60
Taiwan, West Japan	7,507	180–230	120–150	
Marianas	11,034	170–220		
East Japan	9,810	170–320	100–110	38
Kurile	9,783	200–260	90–105	34
Kamchatka		300–340	90–95	34
Aleutian	7,679	150–220		28
Alaska			80	
Mexico		240–360	115–125	39
Middle America	6,662	190–210	115	
Caribbean				
(West Indies)	9,200			
Colombia,				
Ecuador, Peru				22
Chile	8,055	260–320		23
Scotia (South				
Antilles)	8,264			
West Antarctica				

[1] Fisher and Hess, 1963, except for the Kurile Trench (Nitani and Imayoshi, 1963).

[2] Suzuki, 1966.

[3] Dickinson and Hatherton, 1967.

[4] Benioff, 1954.

[5] T = tholeiite magma; H = high-alumina basalt magma; A = alkaline basalt magma (Kuno, 1966).

[6] Holmes, 1965, except the data for Marianas and east Japan, which are from Kuno (1962).

[7] In 10^{23} erg per 68 years per 1° in length of arcs (Duda, 1965).

[8] In numbers per year per 1,000 km in length of arcs (Sugimura, 1967).

Dip of seismic plane for deep foci[4]	Petrographic province for Quaternary volcanoes[5]	Number of active volcanoes[6]	Energy of shallow earthquakes[7]	Frequency of intermediate and deep earthquakes[8]
58,64	H–A	5	} 3.64	} 0.19
	T	12		
	H	12	} 5.02	} 0.26
61	H	19		
	H–A	60		} 0.06
	H–A	12		
60	H	6	} 7.32	} 0.07
		12		0.05
	H–A	16		
75	T	6	0.51	} 0.09
	T–H–A	39		
	T–H	39	} 15.88	
58	T–H–A	25		} 0.17
	H–A	18	} 6.45	} 0.05
	H–A	15		
	H–A	12	} 4.15	} 0.08
	H	31		
		8		0.00
47		8		} 0.11
58	T	26	} 8.60	
	T	2		} 0.01
	H–A	7		

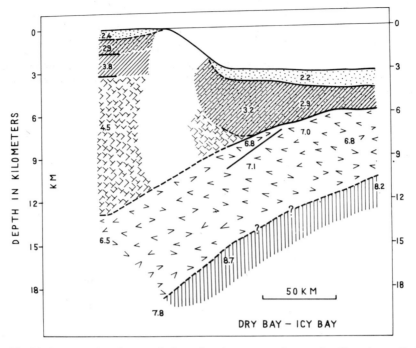

Fig.56. Structure based on explosion seismology across the trench off southeast Alaska. The patterns are adapted to Fig.54. (After Shor, 1966.)

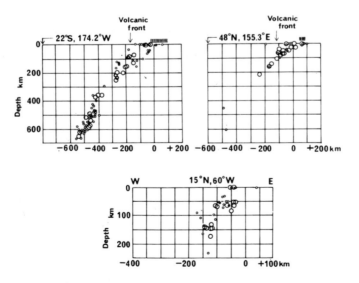

Fig.57. Distribution of foci of intermediate and deep earthquakes in a vertical plane perpendicular to the arc-axis, for the Tonga—Kermadec, Kurile and West Indies arcs. (After Sykes and Ewing, 1965: Sykes, 1966.)

and Sykes, 1969; Santo, 1969a, b, 1970; Fitch and Molnar, 1970; Ishida, 1970) on various arcs. Some of their results are reproduced in Fig.57. According to Sykes, and Ishida, the seismic plane can be defined to a thickness of 50—100 km in the Tonga—Fiji area and the Japanese area where the data are most abundant and in places the thickness may be even less than 20 km. In the past, the notion of "seismic plane" was considered to be a somewhat idealized one for ill-defined groups of earthquakes but now the notion seems to have a definite physical reality. One remarkable finding of Sykes (1966) may be the fact that the sharp horizontal bending of the Tonga Trench at its northern end is reflected by the similar bend of the deep seismic plane. In Fig.58 epicenters of shocks deeper than 500 km and active volcanoes are shown. The parallel arrangement of trench, volcanoes and deep earthquakes is most impressive. Sykes et al. (1969) recently found that the zone of deep earthquakes can be well defined by earthquakes with a magnitude as small as 3. There are, on the other hand, certain areas where the seismic plane can hardly be defined even with the new redeterminations, especially when the deepest shocks occur only at intermediate depths, such as the northwestern portion of the Indonesian Arc (Santo, 1969b).

It has been shown already (see p.40) that, according to the focal mechanism studies, the direction of the axis of the maximum compressive stress causing deep earthquakes under the Japanese arcs is predominantly parallel to the dip of the "seismic plane". This

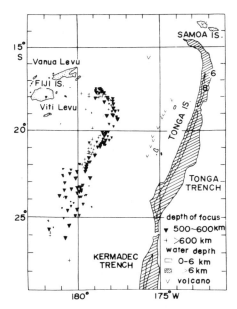

Fig.58. Distribution of trenches, volcanoes and epicenters of earthquakes deeper than 500 km. (After Sykes, 1966.)

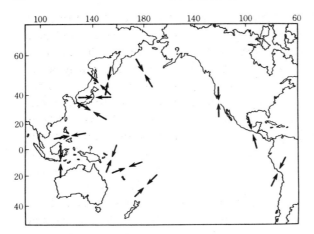

Fig.59. Directions of maximum compressional stress in horizontal plane estimated from the first motions of earthquakes. (After Ritsema, 1961.)

seems to hold for other island arcs as well (Fig.130, p.198). The direction of the maximum stress axis in the horizontal plane has long been noted to be perpendicular to the arcs as shown in Fig.59 (Ritsema, 1961). The marked obliqueness that can be seen in the western coast of the North American continent seems to reflect the non-arc but transform fault characteristics of the San Andreas fault. McKenzie and Parker (1967) examined the directions of slip-vectors of earthquakes in the north Pacific margin from California to Japan, obtained mainly by Stauder and colleagues (Stauder and Udias, 1963; Udias and Stauder, 1964; Stauder and Bollinger, 1964, 1966), and indicated that the slip directions are consistent with the premises of plate tectonics; the Pacific plate and the plate containing North America and Kamchatka are in relative rotation.

Geographical coincidence between the epicenters of shocks with the focal depth of 100–250 km (Fig.27) and active volcanoes (Fig.38) was noted in Japan a long time ago (Honda, 1934a, Wadati, 1935). The same tendency can be also observed in other arcs. Fig.51, 57 and 60 illustrate the cases for the Indonesia, Kurile, Tonga and South America areas. Table IV lists the depth, h, of earthquakes occurring under the front of active volcanoes for various arcs (Dickinson and Hatherton, 1967) and the horizontal distance, d, between the front of active volcanoes and the trench axis (Suzuki, 1966). If we assume that the trench is the expression of the intersection between the seismic plane, dipping with the angle θ, and the earth surface, the relation $d = h \cot \theta$ should hold. In Table IV, θ (Benioff, 1954), d and h as determined by different investigators are listed. These independently determined values seem to satisfy, although roughly, the above relation.

According to Kuno (1966), zonal arrangement of the composition of basaltic magmas (see p.59) also applies to other circum-Pacific areas as shown in Fig.61. Appearance of the three types of magmas, which are called tholeiite, high alumina and alkaline basalts in the order of increasing value of $(Na_2O + K_2O)/SiO_2$, are always in the same order.

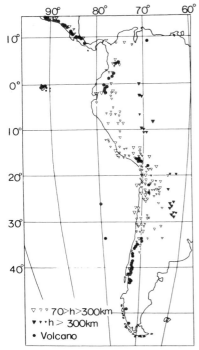

Fig.60. Epicenters of intermediate and deep earthquakes and volcanoes in South America. (After Gutenberg and Richter, 1954.)

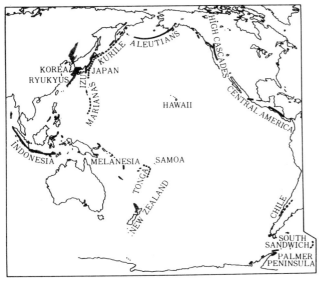

Fig.61. Distribution of three types of basaltic magmas. (After Kuno, 1966.) Broken line = tholeiite; solid line = high-alumina basalt; hatches = alkali-basalt.

However, there are exceptions: some arcs lack tholeiitic basalt whereas some produce tholeiitic basalt only, as listed in the sixth column of Table IV. It may be of interest to note that the kind of magmas and the focal depth of deep earthquakes are related (see p.182). This correlation indicates the possible causal relation between deep shocks and magma production: the more silica-rich and the more alkali-poor basaltic magmas are produced at relatively small depth, whereas the silica-poor and alkali-rich magmas are produced at greater depth, provided that magmas are generated at the deep earthquake foci (Kuno, 1959). In the general scheme of the global tectonics western North America now belongs to the rift and associated transform fault system. Therefore, the petrographic province of this area may seem to require studies as a part of that of the mid-oceanic ridge system which is entirely missing in Fig.61. But, at the same time, as has been suggested by Hamilton (1969), McKenzie and Morgan (1969) and Atwater (1970), western North America was probably an active arc with a deep earthquake zone and volcanism until sometime in the Tertiary Period. Thus it would be natural to discover a zonal arrangement similar to that of the present arcs in western North America. One is, in such a case, dealing with a fossil or paleo-island arc.

In connection with the correlation between the positions of active volcanoes and those of deep earthquakes, one rather interesting point may be worth noticing: the phenomenon which may be called a kind of complementary or exclusive relation between volcanic and deep-seismic activities. On a global scale, the coexistence of deep earthquakes and volcanoes is a characteristic of the inner belt of island arcs. But if one compares Fig.26 and Fig.38, it may be noticed that on a more local scale, the area where active volcanoes are most densely distributed is devoid of epicenters. The same tendency may be observed in Fig.60 for South America. Coexistence on a larger scale and exclusiveness on a smaller scale between volcanism and seismicity may seem incompatible at first sight. In fact, the reality of the latter nature should be checked more carefully before attempting to explain it by a hypothesis such as the one favoured by the present authors, i.e., production of magma needs heating by deep seismicity, but when magma is really produced at a spot there would be no more seismicity because of the easily flowing property of the molten magma.

The trench area of the descending lithosphere should show low heat flow. This is expected from the relatively low temperature of the lithosphere as proposed from its higher seismic wave velocity and lower attenuation, as well as from the general property of the convection cell system that the descending part has least output of heat. Heat flow measurements in island arc areas have been made, besides in the Japanese and Kurile arcs, in the Indian Ocean off Sumatra (Vacquier and Taylor, 1966), at one locality in Chile (Diment et al., 1965) and at several stations off its coast (von Herzen and Uyeda, 1963), at about ten stations in the Aleutian Trench and the Bering Sea (Foster, 1963; T. Watanabe, personal communication, 1966), and at some 30 stations in the seas around the West Indies (see Lee and Uyeda, 1965). With regard to South America, in 1969 a heat flow survey was commenced by the Earthquake Research Institute, University of Tokyo with the cooperation of South American scientists. Their preliminary results indicated low heat flow in the coastal zone and variable heat flow in the Andean zone (Uyeda and Watanabe, 1970). In 1967, an extensive heat flow survey was made in the Melanesian seas by the Scripps Institution of Oceanography in cooperation with Japanese scientists (Sclater and Menard, 1967; M. Yasui, personal communication, 1967). Their results are summarized in Fig.62. In the New Hebrides Arc, the trench is on the western side and the descending slab is considered to dip toward the Pacific Basin. The "reversed" island arc region appears to have a marked high heat flow area in the basin behind the arc (north Fiji Basin). But such a persistent high heat flow area is not observed in the Tonga Arc region (south Fiji Basin). Although McKenzie and Sclater (1968) generalize the high heat flow in the inner zones of the arcs of the western Pacific, the situations of arcs other than the Kurile, Japan and Ryukyu arcs are not yet so clear.

The complementary or exclusive relation between volcanism and deep seismicity mentioned above may be extended to a similar relation between heat flow and deep seismicity (see p.65). In fact, according to Yasui et al. (1968b), such a relation seems to exist in the seas of Okhotsk and Japan. Under the inner zone of the Tonga Arc where the deep seismicity is extremely active, the upper mantle may be not "yet" hot enough, because the surface heat flow in the south Fiji Basin is not uniformly high (Fig.62). In

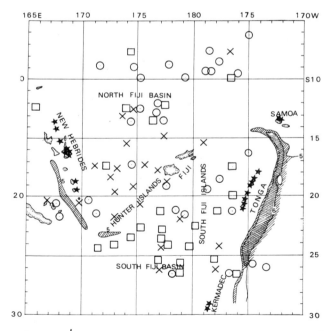

Fig.62. Distribution of heat flow in the Fiji basins. X = heat flow >2.0 H.F.U.; O = 1.0 H.F.U. $<$heat flow <2.0 H.F.U.; $\square$ = heat flow <1.0 H.F.U. Stars indicate volcanoes. (After Sclater and Menard, 1967; and M. Yasui, personal communication, 1967.)

Fig.53 (Oliver and Isacks, 1967), the upper mantle under the Fiji Plateau is shown to be of high Q.

Taking these complex factors in account, it can still be said that heat flow is uniformly subnormal in the western Pacific Basin, and that the heat flow in the "inner" zones of island arcs is variable and often high. It can be expected that the electrical conductivity of the mantle (see p.70) is one of the phenomena directly controlled by the temperature distribution, and heterogeneous temperature should give rise to an anomaly of the conductivity (Uyeda and Rikitake, 1970). Japanese arcs indeed show such an anomaly. As for other arcs, only the South American Arc has been studied from the standpoint of the conductivity anomaly (e.g., Forbush et al., 1967). It was found that the South American coast and the Andean zone have the characteristics which are quite different from those found at non-island-arc type coasts such as the west coast of North America (Schmucker, 1964) and Australia (Parkinson, 1964). This observation seems to harmonize with the expected difference in the temperature distributions in the upper mantle under island arc and non-island-arc type continental margins. We will discuss this problem more fully in Chapter 3.

CHAPTER 2

History of Island Arcs

Attention is directed in this chapter to the geological history of the Japanese Islands and its relation to the island arc features. One of the most important points in discussing the history is the taking into account of the order of length of the period concerned. For the Japanese Islands, there are three orders of critical time span, the longest being more than 300 m.y., the intermediate 26 m.y., and the shortest about 2 m.y. The oldest rocks so far known in the Japanese Islands are of 1,500–1,700 m.y. ago and the oldest of the Silurian Period, about 400 m.y. ago. The geological history of Japan covers more than 1,000 m.y. The last 26 m.y. of the history is the "island arc age". The geographical distribution of island arcs has not undergone any striking change during the 26 m.y. since the earliest Miocene Epoch. But, most of the present island arc features are suggested to have become remarkable during the Quaternary Period, at most since 2 or 3 m.y. ago. The present major relief of the island arcs and the trenches appears to be essentially the result of the recent tectonic movements during the last 2 m.y.

A full description of the geological history of Japan requires a thick monograph by itself. The Japanese Islands are composed of several orogenic belts, like most continental areas. For convenience of description, we have followed the classification of the geological ages given by Ichikawa (1958, 1964):

(1) Archeotectonic age: pre-Silurian (until 420 m.y. ago);
(2) Paleotectonic age: from Silurian to earliest Mesozoic (until 200 m.y. ago);
(3) Mesotectonic age: most of the Mesozoic and Paleogene (until 26 m.y. ago);
(4) Neotectonic age: Neogene and Quaternary.

We have already some evidence, although it is partly fragmentary, to construct the pre-Cenozoic history of the Japanese area, and the authors do not regard it as impossible to do so. However, they are afraid of the present ignorance concerning some information of key importance, for instance, on the drift of continents and the existence of mid-oceanic ridges around Japan during remote geological periods. Since the present island arc features seem to have made their appearance in the beginning of the neotectonic age, the history of the preceding ages will be mostly omitted. In this chapter firstly the existence of the youngest orogeny of the "island arc age" distinguishable from the pre-Miocene orogenies will be focussed on, instead of dealing with the whole geological history of Japan.

Their history since the earliest Miocene will be described in the first half of this chapter and the more substantial discussion about the relation with the older orogenic belts will be included in this part.

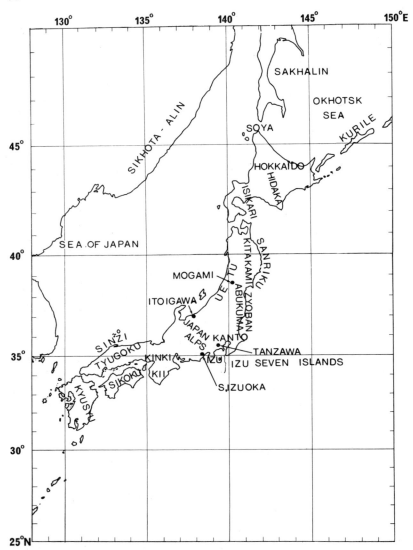

Fig.63. Index map for Chapter 2.

Although the Japanese Islands have experienced no drastic change from the earliest Miocene to the present, the tectonic and igneous activities have not been uniform in time and space. To realize the topographical and geophysical features of the present island arcs, a special attention is given to the Quaternary Period in the second half of this chapter, because the various activities seem to have continued during the Quaternary Period essentially as they are doing now.

SEQUENCE OF OROGENIES IN JAPAN

Main orogeny and Hidaka orogeny

Throughout the paleotectonic and mesotectonic ages, an orogeny or orogenies took place in Japan and formed the framework of the ancient Japanese Islands. The main orogeny referred to here covers most parts of the three larger islands, Honsyu (Fig.10), Sikoku and Kyusyu, (Fig.63), and represents its structure typically in southwest Japan (*IIIA—E* in Fig.36), including the Ryoke and Sanbagawa metamorphic belts (Fig.37). In this orogenic belt an increasing number of phases of movement and even of orogenies have been recognized, dating from the Permian to the Cretaceous. Despite the complexity in time, the resulting geological trends combine into a parallel meandering pattern of apparent simplicity (Fig.75, p.114). Had it not been for the later horizontal deformation (see p.74), these trends would probably have had a regular arcuate shape when the geosyncline was born and when the orogenic movement was active. The history of these times is omitted here for the reason mentioned in the preface of Chapter 2, but the last two remarkable events of the main orogeny are described, which occurred in the Late Cretaceous and the Paleogene. One is the batholithic invasion and volcanic emergence of acidic magmas along the Hirosima belt (*3c* in Fig.65, p.99); the other is the thick non-fossiliferous sedimentation of strata along the Simanto belt (*3a* in Fig.65). In Chapter 3 (see p.202), these two events will be taken as the basic processes characteristic to the Pacific-type orogeny.

The granitic rocks exposed in Japan are almost entirely of this period, the exposure occupying about 15% of the area of Japan. The batholiths were formed by several successive intrusions in the extensive areas especially in Tyugoku. They appear to represent the eastern border of the late mesotectonic igneous activity in the east Asia region, which includes the intrusion of Yenshan granites scattered over more than 4,000 km from south China to northeast Siberia (Fig.133, p.203).

There is a flysch sequence at least 10,000 m thick in the outer Simanto belt, which appears to the south of the Hirosima belt and is separated from this by the median belt (*3b* in Fig.65), where neither igneous activity nor marine sedimentation occurred during the latest Cretaceous and entire Paleogene periods. It seems likely that the strata in the Simanto belt are geosynclinal accumulations deposited largely during the upwarping period of the median belt. Deposition of the flysch sequence of the Simanto Group was extremely scanty in fossils and characterized mainly by alternation of shale and sandstone with graded bedding interbedded intermittently with conglomerate, limestone, radiolarian chert, etc. The rocks are turbidites that are considered to have accumulated somewhere at the foot of a "continental" slope (Kimura, 1966). It was recently inferred, on the basis of the direction of sole markings, that off the southern coast of southwest Japan there was formerly a land which has now vanished. Moreover, judging from the occurrence of orthoquartzite pebbles in the conglomerates of the Simanto Group the land that vanished had a continental crust. This provides one of the major problems and will be discussed later (see p.120). The sediments of the Simanto Group were highly deformed probably in the Early Miocene Epoch. This deformation will be described in the section on the Late Cenozoic history (p.111).

The Hidaka orogeny in Hokkaido (*IIA−D* in Fig.36) is regarded as having the same age as that of the Alpine orogeny in the type region (Hunahashi, 1957). That is, the geosynclinal stage of this orogeny was of the Jurassic Period, and after that time the strata were metamorphosed to form a pair of the metamorphic belts (Fig.37): the high-p/t type Kamuikotan belt (*2b* in Fig.65, p.99) and the low-p/t type Hidaka belt (*2c*). Later during the Tertiary Period, especially in the Miocene Epoch, an intense crustal deformation occurred along the belt (*2a*) that runs in the meridional direction from the western part of Sakhalin through the Soya Strait to the western part of the central range of Hokkaido. As a result of this tectonic movement complicated *Decken* structures were formed in some areas in this belt (Nagao, 1933). The folded belt, occupying the western strip of the Hidaka orogenic belt, was probably a small geosyncline in front of the Hidaka range, and corresponds to the Molasse trough or the peri-Alps depressions in the Alpine orogeny.

For the main orogeny, the high-p/t type metamorphic belt lies on the oceanic side of the low-p/t type belt (see p.53) and a later geosynclinal depression extends along the oceanic side of these belts. The same relation is found in other parts of the circum-Pacific metamorphic belt. One of the remarkable facts observed in the circum-Pacific paired metamorphic belts is that the low-temperature metamorphic belt lies always on the Pacific side and the high-temperature metamorphic belt on the continental side. However, for the Hidaka orogeny of Hokkaido, the low-p/t type belt lies on the east of the high-p/t type belt and the depression of the later stage lies to west: these metamorphic belts do not extend along the present Pacific coast, and the usual "oceanic side" is toward the Sea of Japan, and the continental side is toward the Sea of Okhotsk (Fig.37). The case at a glance may appear to be exceptional.

A hypothesis states that the Hidaka belt once extended approximately from west to east and later rotated clockwise (Matsuda and Uyeda, 1971; Kawai and Nakajima, 1970). According to this hypothesis, the "oceanic side" was literally facing the Pacific at the time of formation. But it is better to interpret it otherwise. Present Melanesian island arcs lie with their backs to the central part of the Pacific and the "oceanic side" or the trench side faces the Australian continent (see p.76). It should be noticed here that there is a deep sea reaching 4,716 m in depth between the Melanesian island arcs and the Australian continent. The floor of the deep sea must consist of an oceanic crust. A similar relation stands with the northeastern part of the Sea of Japan. The sea reaches 3,669 m in depth and its floor is also estimated to consist of an oceanic crust or a transitional type of the oceanic crust. It is suggested here from these relations that the Hidaka belt was formed with a close relationship with the oceanic part of the floor of the Sea of Japan and the land mass of eastern Hokkaido. In this view, the western high-p/t type metamorphic belt was the sea-side feature and the eastern low-p/t type metamorphic belt was the land-side feature.

Neotectonic age

The neotectonic age, the Neogene and Quaternary periods, greatly differs from the Paleogene Period in both the amount and distribution of sedimentary and volcanic rocks in Japan. The thick strata of the Neogene Period are found throughout the Japanese Islands, whereas most parts of the islands were emerged and denuded during the Paleogene. The remaining part, where the sediments accumulated during this time, occupies rather small areas except for Hokkaido and northern Kyusyu, that is, the north-eastern and southwestern ends of Japan. Between them the Paleogene strata appear in narrow strips along the Simanto geosyncline in west Japan (Fig.65, p.99) and only in small patches along the Pacific coast in east Japan. On the other hand, the Neogene and Quaternary strata show an extensive distribution (Fig.69, p.103).

There are remarkable differences also between the volcanic products of the two periods. The distributions of volcanic products in three periods during the Neogene and the Quaternary present similarities to each other (Sugimura et al., 1963, fig.6–8), whereas a distinct difference in distribution is recognized between the Paleogene (Sugimura et al., 1963, fig.5) and the later three periods. The Paleogene volcanic products are scarcely embedded in the deposits in a few limited places along the Pacific coast. Furthermore, the Neogene and Quaternary periods display not only violent volcanic activity but also intense earth movements. Both the distribution of the sediments and volcanic products and the general trend of the earth movements in these periods include a characteristic zone which extends from Kurile through northeast Honsyu to the Izu Seven Islands. This trend has existed since the beginning of the Neogene and forms the present east Japan island arcs. There was a drastic change in the crustal evolution from the Late Oligocene to the Early Miocene.

From the paleogeographical study it is also known that the pattern of shoreline distribution of the Japanese Islands changed discontinuously during the period from the end of the Oligocene to the beginning of the Miocene (Otuka, 1939). One of the paleogeographical maps of northeast Japan in the Early Miocene is shown in Fig.64. In this map, A indicates the area of open sea environment, B the area of coastal environment, and C the land area. Evidence of this paleogeography comes mainly from the distribution of strata and the characteristics of fossils. The sea extends inland through northeast Japan, where a minor archipelago can be imagined, and inversely the lands are distributed over the present sea area. The change of shoreline in Early Miocene time would imply the birth of a minor geosyncline, which extends not along the "Honsyu Arc" but from the Northeast Honsyu Arc to the Izu–Mariana Arc. The geosyncline received a continuous succession of one-cycle sediments (terrestrial, marine and again terrestrial) from the Miocene to the

Fig.64. Paleogeographical map of part of Japan in the Daizima age (ca. 25 m.y. ago) of the Miocene Epoch. *1* = locality where *Lepidocyclina* occurs; *2* = locality where Aniai fossil flora occurs; *3* = locality where Daizima fossil flora occurs; *4* = locality where *Pecten*–Brachiopoda–*Balanus* fossil fauna occurs; *5* = locality where *Vicarya–Cyclina–Soletellina* fossil fauna occurs; *A* = zone where open-sea environment predominated; *B* = zone where embayment or fresh-water environment predominated; and *C* = estimated land area. (After Fujita, 1960.) The outer limit of *C* is unknown, remaining a possibility of a group of only small islands for the Sea of Japan side.

Quaternary, as represented by the Late Cenozoic sequence developed in Uetu district. Part of this sequence of sediments contains petroleum.

Late Cenozoic orogeny and island arcs

The birth of the geosyncline and the sedimentation (see Fig.71, p.103) are almost simultaneous with and closely related to the occurrence of abundant volcanic products in the Lower Miocene terrestrial and marine sediments. The volcanic products are generally called "Green Tuff". The volcanic activity was high during the Early and Middle Miocene. Later, the belt (*A* and *B* in Fig.64) was developed into a folded and faulted zone (see Fig.78, 79, pp.120–124). Embryonic folding and faulting are recognized to have already commenced soon after the sedimentation, but the climax of these movements began at the close of the Tertiary Period when the sedimentation tended to terminate. In the belt, granitic rocks intruding into Neogene strata are known. The plutonism is of the Middle Miocene in age according to the biostratigraphic evidence as well as the radiometric age determinations. Metamorphic rocks formed in this age are also known from some areas in the belt (e.g., Seki et al., 1969).

The volcanic activity, tectonic movements and plutonism, that have taken place along the east Japan island arcs since the beginning of the Neogene, are attributed to one process, which should be called an orogeny because it has almost every aspect of the orogenic movement and yet it is apparently independent of any preceding orogenies in the area. The geosynclinal axis as well as the centre of the volcanic belt of this period neither follows nor parallels those of the preceding orogenic belts, but crosses the older structures in two or three areas: in the northeast it crosses the Hidaka Range to extend to the inner side or the Sea of Okhotsk side of the Kurile Arc and in the south it crosses the Japanese main orogenic belt to extend itself to the Izu–Mariana Arc.

The younger belt is thus considered to have become an independent orogenic belt since the Miocene, the cycle of which belongs to neither the Alpine orogeny nor the Japanese main orogeny. For these reasons Sugimura et al. (1963) named this cycle "Mizuho orogeny", Mizuho being an ancient name for Japan. This orogeny includes both the "Green Tuff movement" and the contemporaneous movements around the Japan Trench. The Mizuho orogeny was summarized by Matsuda et al. (1967).

As the continuation of the Mizuho orogeny, during the Quaternary Period, the geosynclinal belts have been folded, uplifted, and covered by volcanic products. The Quaternary volcanoes forming the east Japan volcanic belt (see p.55) also extend southward to the Marianas and northward to the Kurile. The present island arc features are nothing but the manifestations of the Mizuho orogeny. The geographical extension of such movements as folding, faulting, warping, and volcanic activity in the Late Cenozoic age coincides almost exactly with the inner side of the island arcs delimited by the present phenomena. This fact is exemplified by the comparison of distributions of the recent volcanoes (Fig.38) and the Early Miocene volcanic rocks (Fig.66, p.100). The Mizuho orogeny, thus, seems to be still in progress. Finally, we emphasize that the younger orogenic belt should not be limited to the land area but should include the sea-floor extending to the Japan Trench.

It seems to be significant to compare our view on the Late Cenozoic orogeny with the view of M. Minato (in Minato et al., 1965), who first emphasized the existence of a new orogeny independent of the Alpine orogenic cycle (Minato, 1952). The points of agreement are as follows:

(1) At the beginning of the Miocene, the Japanese Islands reached the present island arc stage, showing a clear contrast with the pre-Miocene quiescence.

(2) The structure of the new orogeny is not parallel to the older structure: the basic trend of the present island arcs is oblique to that of the older orogeny.

(3) Along the island arcs an echelon structure is well developed.

Our view differs from Minato's in the following point: he emphasizes the formation of many large faults in the Early Miocene along the Green Tuff region, which were the cause of the intensive volcanic activity and regional subsidence. But we find no evidence for the Early Miocene faulting on such a large scale.

At least three orogenies can be distinguished in Japan. They show the different trendings with each other. If we go far back from the present, the orogenic features show increasing obscurity. The last orogeny has taken place along the east Japan island arcs as well as the west Japan island arcs. The fundamental framework of the present east Japan island arc features originated at the beginning of the Neogene, and the same endogenetic agency has continued to act, with some variations, up to the present. Most of the tectonic and volcanic processes and events during the neotectonic age in east Japan appear to belong to one endogenic series, the Mizuho orogeny. The recent island arc features in east Japan can be regarded as an expression of this orogeny. Those in west Japan may also be interpreted in the same way.

LATE CENOZOIC HISTORY OF JAPAN

The general outline of the framework of Japan based on the neotectonic point of view is shown in Fig.65 which demonstrates the structural division of the islands. Numerals with a letter such as *1a*, *1b*, and so on appearing in parentheses in this section indicate the divisions in the figure.

The Mizuho orogeny is the most recent orogeny in Japan, and the orogenic belt lies along the Kurile Arc, the Northeast Honsyu Arc, and the Izu Mariana Arc, which constitute the east Japan arc system. The Northeast Honsyu Arc is divided into three belts of different structures and histories. They are, from the oceanic side to the continental side, the Off–Sanriku–Zyoban geosyncline (*1a*), the Kitakami–Abukuma geanticline (*1b*), and the Uetu–Fossa Magna idiogeosyncline (*1c*). The name of the Off–Sanriku–Zyoban geosyncline was proposed originally by Matsuda et al. (1966) and will be defined for the first time on p.107 in this monograph. Its concept is important from the standpoint of the present authors.

In the earliest Miocene (Aquitanian), the area of the Uetu–Fossa Magna geosyncline suddenly became a basin filled with sediments, and then a region of violent submarine, but partly subaerial, volcanism, followed by a regional subsidence in the Early Miocene (Aquitanian to Burdigalian). Neither volcanic activity nor regional subsidence had occurred in the same region before the earliest Miocene, at least not during Paleogene time (Minato et al., 1956).

Miocene volcanism

Early in the Miocene Epoch, in the most part of the Uetu–Fossa Magna geosyncline and the Sea of Japan coast region of southwest Honsyu (northern strip of *3c*) an intense volcanic activity associated with fairly rapid subsidence started. It can be suggested by the

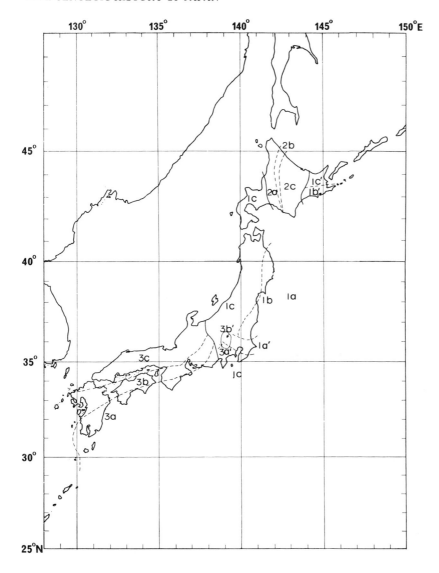

Fig.65. Tectonic division of the Japanese Islands. $1a$ = Off–Sanriku–Zyoban geosyncline; $1a'$ = Kanto structural basin; $1b$ = Kitakami-Abukuma geanticline; $1c$ = Uetu–Fossa Magna idiogeosyncline; $1b'$ + $1c'$ = east Hokkaido province; $2a$ = Yezo folded belt; $2b$ = Kamuikotan high-p/t metamorphic belt; $2c$ Hidaka low-p/t metamorphic belt; $3a$ = Simanto geosyncline; $3b$ = median geanticlinal belt (southern strip: Titibu unmetamorphosed belt; central strip: Sanbagawa high-p/t metamorphic belt; and northern strip: Ryoke low-p/t metamorphic belt; see also Fig.36). $3c$ = Hirosima acidic igneous province; and $3a'$ + $3b'$ = Kanto Range. These divisions would be pertinent for the neotectonic age and the later half of the mesotectonic age.

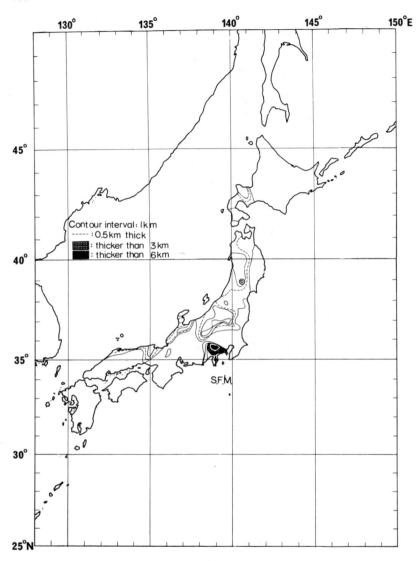

Fig. 66. Thickness of volcanic products of the Early and Middle Miocene Epoch. (After Sugimura et al., 1963.) S.F.M. = South Fossa Magna.

sequence of volcanic products that in the beginning there were subaerial eruptions, but later the sea invaded the subsided areas, and the whole area came to be occupied by submarine volcanoes (Fig.66). The volcanic products in this first period of the Late Cenozoic orogeny range from basic to acidic in composition, and most of them have been altered and are characterized by epidote, chlorite and other clay minerals showing a greenish colour, after which they are often called "Green Tuff" as already mentioned.

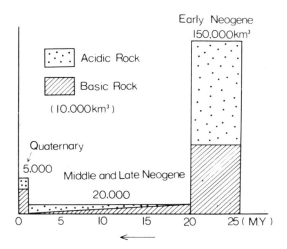

Fig.67. Histogram showing the change in volume of volcanic products during the Neogene and Quaternary periods. The far-right part shows a very small amount for the Paleogene Period. (After Sugimura et al., 1963.) If we follow the dates estimated by Chinzei (1967) as in the caption of Fig.70, the age dividing into the Quaternary and Neogene should be 2 m.y. ago and the age dividing into the Middle and Early Neogene should be 22 m.y. ago, but the general tendency of the change in volume will remain unchanged.

Volcanic activity at the time, as seen in Fig.67, was more violent than at present, and the production of lavas and other effusives per unit time reached five or six times more than that of the Quaternary Period inclusive of the present time. Although in this figure the volcanic activity seems to have decreased abruptly at the end of the Early Neogene, it is only an apparent feature due to the long time interval of taking sums. According to a more detailed but local estimate by Ozawa (1968), shown in Fig.68, the volcanic activity was most intense at the beginning of the Early Neogene and decreased gradually with time. (The beginning of the Onnagawa Age in Fig.68 corresponds to the end of the Early Neogene in Fig.67.)

The Uetu—Fossa Magna idiogeosyncline

The thickest strata of the Japanese Neogene System are in the Uetu—Fossa Magna geosyncline (Fig.69). They are up from a few to ten kilometres thick, including the above-mentioned basal volcanic products (Fig.66). The growing process of the formations is well illustrated in Fig.70. Correlation with radiometric ages by the use of planktonic Foraminifera was attempted and a chronological table of stratigraphic sequences from seven districts in this area was made by Chinzei (1967). He also prepared a diagram showing the present relief and the Late Cenozoic history, across the geosyncline and the Kitakami Mountains in the east (Fig.71). The reader may realize, by observing this diagram from bottom to top, that the geosyncline, shown in the left-hand side of this

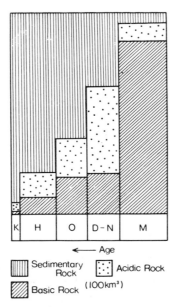

<— Age

| Sedimentary Rock (image) | Sedimentary Rock | Acidic Rock (image) | Acidic Rock |
| Basic Rock (image) | Basic Rock | (100km³) | |

Fig.68. Diagram showing the change in ratio between the volume of sedimentary rocks and that of volcanic products during the Early and Middle Neogene Epoch in some area in northeast Japan. (After Ozawa, 1963.) The length of abscissa is in proportion not to time, but to total volume. K = the earlier half of the Kitaura Age; H = Hunakawa Age; O = Onnagawa Age; $D-N$ = Daizima−Nisikurosawa Ages; and M = Monzen Age. For the ages, see also the caption for Fig.70.

figure, was invaded by the sea, changed to land areas, and then reached the present heights comparable to those of the eastern Kitakami Plateau on the right-hand side of the figure.

Volcanic activity, rapid sedimentation and tectonic movements continued throughout the Miocene in this "idiogeosyncline". However, in the beginning of the Miocene, volcanic activity and subsidence were intense along the eastern strip of the geosyncline, and in the Middle Miocene the centres of subsidence were gradually transferred to the Sea of Japan side, the volcanic products decreased as shown in Fig.68, and the depth of the sea increased, in some parts to 1,000 m, as demonstrated by fossils.

The subsidence in the western side of the geosyncline culminated during the Middle Miocene, after which the sea gradually retreated. This subsidence is represented by a rather uniform marine oil-producing formation, which consists largely of lower silicious shales (Onnagawa Formation) and upper black siltstones (Hunakawa Formation) overlying the former volcanics (Ikebe et al., 1965, 1967). In the same stage through the Sarmatian, the eastern strip of the preceding volcanism and subsidence showed a tendency to upwarping associated with gentle folding of strata. This upwarping was characterized by the volcanism of basalt and dolerite with acidic rocks. Dolerite sheets are found to be interbedded in the shales and dacitic pyroclastics occur along the upwarped

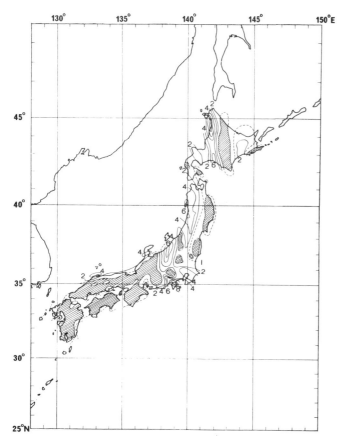

Fig.69. Thickness of sediments deposited since the beginning of the Miocene. Contour interval = 2 km. Hatched areas are where the sediments are only scattered or not present. (After K. Chinzei, personal communication, 1968.)

ranges. The upwarped ranges were arranged en echelon and separated by elongated basins, in which the shales and siltstones were accumulated.

The upwarping was accompanied by acidic plutonism on a small scale (Fig.72). During the Middle Miocene the eastern strip of the geosyncline was invaded several times by granodiorite and quartz diorite and partly metamorphosed mainly in the south of the belt. The largest granitic mass is around Mount Tanigawa-dake, and the "south Fossa Magna" (Fig.66) has the well-studied Tanzawa quartz diorite mass and the larger but less well-known Daibosatu granodiorite north of Mount Huzi (Fuji). More than a score of smaller occurrences are exposed in the Uetu area to the north. Towards the end of the Middle Miocene rapid uplift and faulting took place in the south, and erosion exposed the Tanzawa and other quartz diorites. At the same time a few narrow strips were down-warped in the "south Fossa Magna" and formed subsiding basins in which conglomerate,

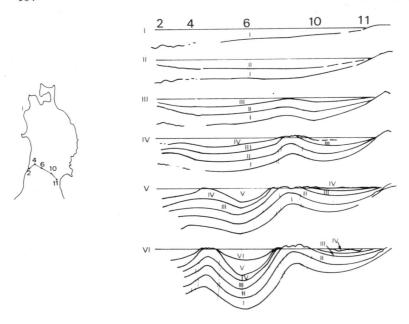

Fig.70. Stratigraphic section showing six stages of the development of the Uetu–Fossa Magna idio-geosyncline. Arabic numerals indicate the locations.

   I = Monzen Age (26.5–26 m.y.);
  II = Daizima Age (26–24 m.y.), and Nisikurosawa Age (24–22 m.y.);
 III = Onnagawa Age (22–17 m.y.);
 IV = Hunakawa Age (17–10 m.y.);
   V = Kitaura Age (10–7 m.y.); and
 VI = Wakimoto Age (7–4 m.y.).

Note that each of these is not a structural section but a restored section, i.e., a continuous series of stratigraphical columns. In *IV, V* and *VI*, some shear planes are suggested. (After Kitamura, 1963; estimated dates are after Chinzei, 1967.)

containing the pebbles of the quartz diorite of later Middle to Late Miocene ages, accumulated. These deposits were faulted and tilted during the Middle and Late Neogene Period.

In most of the Uetu–Fossa Magna geosyncline, the elongated basins became narrower towards the end of the Miocene and diverged into lacustrine basins in the Pliocene. These basins and the intervening ranges have remained in position and have grown in depth and height to the present state (see p.138).

The Kitakami–Abukuma geanticline

The characteristic features of northeast Japan are the two non-volcanic massifs of the Kitakami and Abukuma plateaus (Fig.63). Apart from superficial deposits the two massifs consist mainly of pre-Miocene rocks. Important Neogene System rocks are only exposed

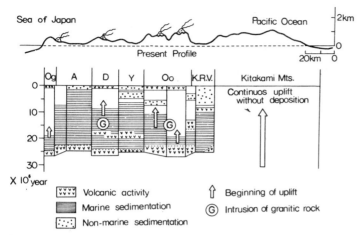

Fig.71. Present profile across northeast Japan from west to east and diagram showing the outline of the history since the beginning of the Miocene. (After K. Chinzei, personal communication, 1968.) *Og* = Oga Peninsula; *A* = Akita Plain; *D* = Dewa Range; *Y* = Yokote Basin; *Oo* = Oou Range; and *K.R.V.* = Kitakami River Valley. Reading the diagram from bottom to top, the development experienced by the left-half Uetu geosyncline from sea to land to the present topography can be seen. (See also Fig.70.)

along a narrow strip of hills between the Abukuma Plateau and the Pacific coast. The total thickness of this Neogene System is about 1,000 m. The strata dip eastward and the structure is simple except for the gentle basin structure and a system of normal faults. The general simplicity of the strata shows that they were deposited along the eastern border of the Abukuma Plateau, that was a fairly stable mass during the course of the Neogene Period. The Abukuma Plateau, bounded on the east by the north—south trending Hutaba fault seems to have been uplifted by about 1,000 m in the Late Cenozoic time (Tsuneishi, 1966). In the northeastern corner of the Kitakami Plateau, recent uplift is proved by the occurrence of raised coastal terraces of the Late Pleistocene Age. This belt of uplift is characterized by a positive gravity anomaly (Fig.13). Matsuda (1964) named this the P-belt.

As in older orogenies there were probably two geosynclines on both sides of this upwarped belt. One of the two geosynclines has been described above as the Uetu—Fossa Magna idiogeosyncline and the other is the Off—Sanriku—Zyoban geosyncline. The latter will be explained in the next section. The geosynclines on both sides of the Kitakami—Abukuma geanticline probably changed the position of the axes of subsidence away from the latter as they sank. It may have been the consequence of the continuous upwarping of the central geanticline.

The updoming of crustal layers of higher wave velocities (Fig.17) in the positive gravity anomaly belt may be related to the uplift of this belt. Miyashiro (1967) gives an explanation for the gravity anomaly which is also consistent with the regional uplift, introducing an upper mantle expansion caused by a phase transition under high-temperature condition. Such a condition is, however, contrary to the observed low heat

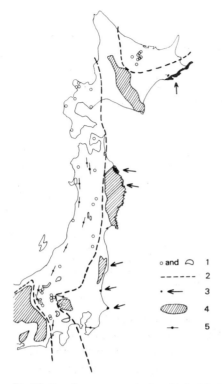

Fig.72. Zonal arrangement of geological features along the Northeast Honsyu Arc.
1 = Middle Miocene acidic intrusive;
2 = Early Miocene volcanic belt ("Green Tuff" area);
3 = Late Cretaceous to Eocene formation not deformed;
4 = continuously emerged area since the Early Miocene; and
5 = direction of Quaternary fold axis.
(After Sugimura, 1962b; for northeastern Hokkaido after M. Shimazu in Minato et al., 1965.)

flow of the positive gravity anomaly belt. The more important matter that Miyashiro has pointed out, may be that the upper part of the crust in the belt including the Kitakami–Abukuma geanticline was eroded away as it rose, at least since the Miocene, and hence the thinner crust than those in the neighbouring areas has resulted in these massifs (Fig.17).

It is concluded that a geanticlinal belt, which has been upwarped or at least has remained stable during the Late Cenozoic, has existed between the Uetu–Fossa Magna idiogeosyncline and the Off–Sanriku–Zyoban geosyncline (see next section).

The Off–Sanriku–Zyoban geosyncline

The parallelism between the youngest active zone, including the Uetu–Fossa Magna geosyncline and the series of trenches – the Kurile, Japan, and Izu–Ogasawara (Bonin) trenches – suggests that the trenches should be an integral part of the Late Cenozoic orogeny. Sugimura (1958) and Sugimura et al. (1963) proposed that the orogenic belt which characterizes the Late Cenozoic activations in northeast Japan should include the area of the Japan Trench and its island side slope. Although this submarine area has not yet been investigated sufficiently well, it seems that the area represents a subsiding belt and has a thick pile of sediments. Such a belt might be called a sort of geosyncline, and may be the major field of the orogeny as a whole. The portion of the belt along the island side of the Japan Trench is named the Off–Sanriku–Zyoban geosyncline. The extent of the geosyncline can be delineated and defined by the negative free-air or isostatic gravity anomaly (Fig.12), provisionally until more information will be obtained. The crustal thickness there has been estimated to be about 25 km from the seismic evidence.

The northerly extension of the Japan Trench negative belt seems to branch off to the Isikari Plain in Hokkaido from the main belt, the latter being parallel to the Kurile Arc. A similar feature is found for the southerly extension. It seems to branch off to the Kanto Plain to the west from the main belt which is parallel to the Izu–Ogasawara (Bonin) Trench. The Kanto structural basin (*Ia'* in Fig.65), which remained submarine from the beginning of the Miocene to the Early Quaternary, probably represents the southwestern branch of this geosyncline. Basin-forming deformation in the Kanto area during the Quaternary will be discussed later (see p.129). The crust under the trench and its island side slope, which coincides with the belt of negative gravity anomaly, must have been depressed. Depression might have occurred at least twice since the Miocene, the last time estimated to be in the Quaternary on the basis of the analogy to the Kanto basin-forming deformation.

Remarkable negative gravity anomalies are found along almost every outer belt of island arcs, including the Indonesian Arc, where the belt covers Timor Island. Molengraaff (1914) showed that the intensely folded Miocene sediments of geosynclinal character occur there. The negative anomaly belt along the Japan Trench may represent similar geosynclinal sediments of younger age, probably of the neotectonic age. They should be intensely folded with the alpino-type structure, if the analogy to Timor Island and the underthrusting of the oceanic lithosphere along the Japan Trench are considered (see p.152). The folding may contrast with that of the Uetu–Fossa Magna idiogeosyncline with a germano-type structure. Besides, it is clear that the latter differs from the Off–Sanriku–Zyoban geosyncline in including abundant volcanic products.

The floor of the Japan Trench has been downwarped to a depth that is probably beyond the reach of supply of terrigenous sediments, except turbidites. Although the deep trench itself may not be a geosyncline of the present day, there is now sufficient general evidence that the sea-floor between the continental slope and the trench is a present-day geosyncline, where the younger sediments may be folded and thrust into the alpino-type structure, which differs from the germano-type structure of the Uetu—Fossa Magna geosyncline. A belt of high seismicity where shallow shocks strongly predominate lies in the area of the Off—Sanriku—Zyoban geosyncline. This seismicity may be related to the present-day orogeny (p.34).

The Japan Trench is the topographic expression of the downwarping, which will be discussed later (p.140), and represents the Quaternary position of the axis of depression, whereas the position of a thick pile of sediments, probably formed in the Miocene Epoch, could have been situated in the middle of the Off—Sanriku—Zyoban geosyncline, the inner side of the Japan Trench. It is likely that the underthrusting is somewhat episodic.

Three belts in East Japan

The Mizuho orogenic belt in northeast Japan, thus, consists of the Uetu—Fossa Magna idiogeosyncline on the inner side, the Off—Sanriku—Zyoban geosyncline on the outer side, and the Kitakami—Abukuma geanticline between them. They are the Inner, Median, and Outer belts in the Pacific-type orogeny framework developed in Chapter 3 (p.200). The characteristic features of the geology are tabulated to clarify the difference between these three belts (Table V). It seems to us that the leading part in the Mizuho orogeny is the formation of the Japan Trench and its extensions and the tectonic movement in the

TABLE V

Geological features of three belts in east Japan

U—F M idiogeosyncline (Inner Belt)	K—A geanticline (Median Belt)	O—S—Z geosyncline (Outer Belt)
Intermediate to acidic volcanism associated with minor plutonism	essentially no volcanism	essentially no volcanism
Block movements resulting in ranges and basins, the latter being extensive in the earlier stage	gentle, continuous upwarping resulting in peneplains	regional subsidence
Moderate germano-type folding and faulting	no folding	probably alpino-type folding and thrusting
The inner arc	the outer arc showing positive anomaly	the outer arc showing negative anomaly and the trench

The bottom of the table shows the corresponding division of the island-arc trench system.

Off—Sanriku—Zyoban geosyncline, whereas the Uetu—Fossa Magna idiogeosyncline and its extensions may be one of the secondary features associated with them.

Since the Early Miocene, the Uetu—Fossa Magna idiogeosyncline has been subjected to active volcanism, which began suddenly after a long period of denudation that had lasted since the end of the Paleozoic. Various volcanic products, mostly acidic to intermediate in composition, were laid down successively in subaerial, lacustrine, and submarine conditions, interbedded by clastic sediments. Granitic intrusions accompanied them. The Quaternary volcanism (see p.55) in the idiogeosyncline is a continuation of the Late Tertiary one. Although Quaternary volcanoes are concentrated in a fairly narrow zone just west of the volcanic front, the Late Cenozoic magmatism is known to have been active over a much wider area, extending to the eastern margin of the main Asiatic continent (Kuno, 1952; Ministry of Geology of the U.S.S.R., 1965). Heat flow is persistently high not only in the volcanic belt on the main islands of Japan, but also over the Sea of Japan (see p.63). Thus, it appears reasonable to consider that the Inner Belt, which will be defined in Chapter 3 as an element of the Pacific-type orogeny, in the Late Cenozoic age of northeast Japan, is broader than the Uetu—Fossa Magna idiogeosyncline, having a width greater than 1,000 km from the volcanic front to the interior of the Asiatic continent beyond the Sea of Japan.

East of the volcanic front lies a median non-volcanic geanticlinal ridge of positive gravity anomaly (Kitakami—Abukuma geanticline) which has uprisen gradually throughout the Cenozoic Era. This geanticlinal belt separates the Inner Belt or the Uetu—Fossa Magna idiogeosyncline from the offshore Outer or Trench Belt.

The Outer Belt is represented, in the present island arc, by the Trench Belt or the Off—Sanriku—Zyoban geosyncline, where the geosynclinal accumulation of sediments is inferred to be occurring. A post-Pliocene large scale subsidence and tectonism of this belt will be suggested later (see p.140). It is inferred that the belt came into existence at least by the Early Miocene, as a counterpart of the Inner Belt.

Underneath the two geosynclines there may be present-day metamorphic belts, which would constitute a pair of low-p/t type and high-p/t type ones characteristic to the circum-Pacific regions (see p.53). Metasedimentary rocks of zeolite facies and greenschist facies observed in the Uetu—Fossa Magna geosyncline (Yoshimura, 1961; Utada, 1965; Shimazu and Sato, 1966) may be the outcrop of the peripheral zone of the buried low-p/t type metamorphic belt. The present-day paired metamorphic belts are expected also in the North Island area of New Zealand (Hattori, 1968).

West Japan

Two arcs are known in west Japan: the Southwest Honsyu Arc and the Ryukyu Arc. The former is discussed first.

The Southwest Honsyu Arc was active in the mesotectonic age, but has not been so active during the neotectonic age, the Late Cenozoic Era. Hence, the area is divided, in

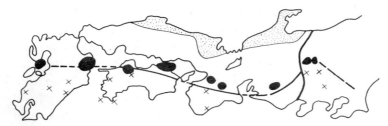

Fig.73. Zonal arrangement of the Miocene igneous rocks along the Southwest Honsyu Arc.
Stippled area = Early Miocene volcanic belt ("Green Tuff" area) after Sugimura et al. (1963); black
area = Middle and Late Miocene volcanoes after Kawachi and Kawachi (1963); cross = Middle and Late
Miocene acidic intrusives after Nozawa (1968); solid line = the Median Tectonic Line parallel to the
coasts and Itoigawa–Sizuoka fault perpendicular to the coasts; and dashed line = supposed extension
of the Median Tectonic Line.

TABLE VI

Miocene igneous rocks in southwest Japan

Age	Northern zone	Median zone	Southern zone
14 ± 3 m.y.	(alkaline rocks)	andesite, dacite, rhyolite and pitchstone of subalkaline magma	granodiorite, granite-porphyry, adamellite, etc.
21 ± 1 m.y.	(calc-alkaline rocks)	–	granodiorite and quartz porphyry
25 ± 1 m.y.	basalt and andesite of tholeiite magma	–	–

The parentheses indicate small amounts.

Fig.65, into three tectonic provinces on the basis of the mesotectonic activity. The
boundaries of these provinces are not effective in the neotectonic age. But the general
trend remains to be the same (Cenozoic Research Group of Southwest Japan, 1960). The
major igneous activities during the neotectonic age are compiled as Fig.73 and Table VI.

The northern zone in the Sea of Japan side (Fig.73) is similar to the Uetu–Fossa
Magna idiogeosyncline in the Miocene history including the "Green Tuff" activity, the
succeeding marine invasion and complicated structures (Matsumoto and Wadatsumi,
1959; Kaseno et al., 1961; Sakamoto, 1966). Although there is a view that the volcanic
province in the northern zone of southwest Japan is a branch of the Lower Miocene
volcanic belt of the Uetu–Fossa Magna idiogeosyncline, the authors interpret that it
may be the volcanic belt produced by the heat generated along the mantle earthquake
zone (see p.177) that resulted from sinking of the oceanic lithosphere in the Nankai
Trough in the Early Miocene Epoch. The intense folding and thrusting, which seem to be

associated with the sinking of the oceanic lithosphere, are shown in the Simanto geosyncline (*3a* in Fig.65) along the Pacific coast of southwest Japan. The Simanto geosyncline developed in the mesotectonic age and continued to subside in the Early Miocene Epoch. The folded strata of the Early Miocene age are unconformably overlain by the Middle Miocene sediments, some of which are gently tilted toward the Pacific and others of which form broad and shallow half-basin structures (Tsuchi, 1961). It indicates that the major deformation of the Simanto belt occurred in Early to Middle Miocene time. Here we refer to it by the name "Takatiho phase" of deformation.

The neotectonic age of the median zone is summarized by Huzita (1962). According to him, southwest Japan shows generally more cratonic characters than northeast Japan during this age. The province subsided in Miocene and Plio-Pleistocene times. During these depressions, thin shallow-sea and terrestrial deposits in the Miocene and lacustrine deposits in the Plio-Pleistocene were formed. Volcanic products abundant in the Middle and Upper Miocene strata are only noticeable geological events in the median zone (Fig.73; Table VI). The peak of the activity was in the Late Miocene Epoch with a second maximum in the Middle Miocene, and the whole activity seems to have ended in the Early Pleistocene Epoch. The long and narrow area of volcanic activity suggests a volcanic belt which, according to Kawachi and Kawachi (1963), extends to central Japan as shown in Fig.73. It is suggested that the volcanic products were derived not from the tholeiite magma but from the subalkaline magma. There was a change in primary magma from tholeiitic in the Early Miocene of the northern zone to subalkaline in the Middle and Late Miocene of the median zone. It appears to indicate that, based on our theory (in Chapter 3), the sinking rate of the oceanic lithosphere underthrusting the southwest Honsyu Arc decreased from the Early Miocene to the Middle Miocene. This decrease seems to correspond to the cessation of the intense deformation in the southern zone at a time between the Early Miocene and the Middle Miocene.

The Late Cenozoic igneous rocks in the southern zone can be divided into two, the ultrabasic to basic rocks in the outer side and the acidic rocks in the inner side. The stocks of ultrabasic composition intruded the Paleogene strata in the Simanto folded belt probably during the Takatiho phase. The ultrabasic and basic activity was slight but may be meaningful for the sinking oceanic lithosphere in the Early Miocene Epoch, as the top of the oceanic lithosphere may have contained a sufficient amount of water to produce serpentinite by the reaction with the peridotite of the continental lithosphere. The heat generation by the shear between the two lithospheric plates (see p.179) may begin to occur at slightly deeper level than that of this reaction. The acidic plutonic and partly effusive rocks are indicated in Fig.73 and Table VI. Most of their dates cluster around 14 m.y. and the remaining at 21 m.y. (Nozawa, 1968). A well understood example, which is, however, an exception to the plutonism, is the volcanic eruption that provided an acid igneous complex of more than 300 km^3 in the Kii Peninsula (Aramaki, 1965; Aramaki and Hada, 1965; Fig.63). The origin of the acidic activity is also suggested to be in the heat energy produced by the sinking of the oceanic lithosphere along the Nankai Trough in the Tertiary Period. Hot springs in the Kii Peninsula and high heat flow in the Nankai Trough may be related to this activity as an after-effect.

To the Ryukyu Arc, we can trace the Takatiho phase folding and the later acidic plutonism, and the Neogene System after this phase shows a relatively simple structure of faulting and gentle tilting. To the northwest, the north Kyusyu area differs from any of the three zones of the Southwest Honsyu Arc. There are Neogene terrestrial sediments conformably overlying the Paleogene strata and the upper part of them is intercalated with the alkaline volcanic rocks.

The areal extent of volcanic activity seems to have retreated from the most part of the Southwest Honsyu Arc to Kyusyu since the Miocene up to the Pleistocene. In recent years, active volcanoes are located only in Kyusyu, from where the west Japan volcanic belt (see p.55) extends to the Ryukyu Arc.

Crossings of orogenic belts

There is a reverse V-shaped syntaxis structure of the Ryoke–Sanbagawa metamorphic belts (Fig.37) and the associated Simanto folded belt, surrounding the Izu Peninsula in central Japan (Fig.74). It is *overlapped* by the north–south trending younger Uetu–Fossa Magna belt. It is interpreted that the mesotectonic structure is inherited by the later tectogenesis: the trends of Miocene depressions follow the older belt rather than the younger one, and the folding axes are not parallel northward to the east Japan arcs but are of reversed V-shape parallel to the extension of the mesotectonic trend of southwest Japan. Some remarks on this crossing of two belts will be given in the later half of this section.

A similar crossing of a younger structure with an older one is found in central Hokkaido, where the north–south trending Hidaka orogenic belt *partitions* the younger belt in a transverse direction into the southwest Hokkaido district to the west and the northeast Hokkaido district (*lc'* in Fig.65) to the east, both of which contain Miocene geosynclinal sediments and volcanic products. The partitioning is so strong that within the central belt of the Hidaka orogeny there is almost no sign of present island arc features parallel to the Kurile Trench.

The third example of such a crossing is *not conspicuous* in appearance but it occurs in the northern district of the Uetu belt (Fig.75). Amid the Neogene strata widespread in the belt, older rocks similar to those in southwest Japan outcrop forming structural islands. They represent the relic of a continuous belt of older rocks that was the extension of the belt of southwest Japan to northeast Japan. This belt, together with that of southwest Japan, was once the backbone of the ancient Japanese Islands formed by the main orogeny. In the northern district of the Uetu belt the younger trend is almost meridional, while the older trend is north-northwest–south-southeast and is largely destroyed by the younger orogeny or concealed under the younger structures.

As italicized above, the expression of older structure is different in the three crossings.

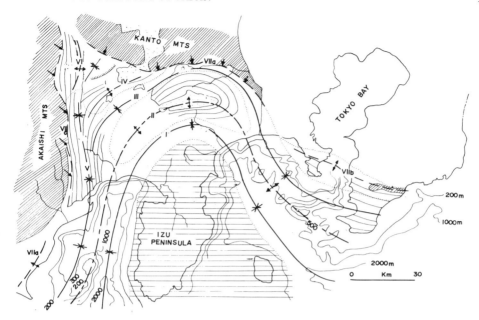

Fig.74. Tectonic map of the south Fossa Magna area. *I— VI* = belts of uplift and subsidence since the Middle Miocene; *VII a, c* = marginal thrusts active since the Early Miocene; and *VII b, d* = marginal belts of uplift since the Early Miocene. (After Matsuda, 1962.)

It is strongest in the case of Hokkaido, and weakest in the Uetu belt. The later the older orogeny survived, the stronger is the expression of its structure at the crossing.

The remaining part of this section is devoted to some discussions on "Fossa Magna" (p.50).

The Median Tectonic Line, which splits the Ryoke—Sanbagawa belt in southwest Japan lengthwise and shows right-lateral movements in the Quaternary Period (Okada, 1968, 1970), terminates in central Japan, where it disappears beneath a cover of Neogene sediments and Quaternary volcanics. The distribution of the cover crossing the central part of Honsyu Island can be clearly seen on the geological map (Fig.36). There the folded sedimentary cover rests on the southwest Japan basement rocks. The ground surface is relatively low except for the volcanic cones, and along the western border of the folded zone of the sedimentary cover the foothills of the Japan Alps appear like a great wall. It is now concluded from the preceeding discussions that this zone is a part of the Late Cenozoic tectonic depression, which extends northward to the Uetu area on the Sea of Japan side of northeast Japan. But it was at first thought by Naumann (1885) that the zone was originated by a great tensional fissure. He was probably impressed by the line of fault scarps and the group of recent volcanoes, and called this folded zone Fossa Magna. According to his interpretation, Fossa Magna is a broad belt of depression caused by a fissure crossing Honsyu Island from the Pacific coast to the Sea of Japan coast,

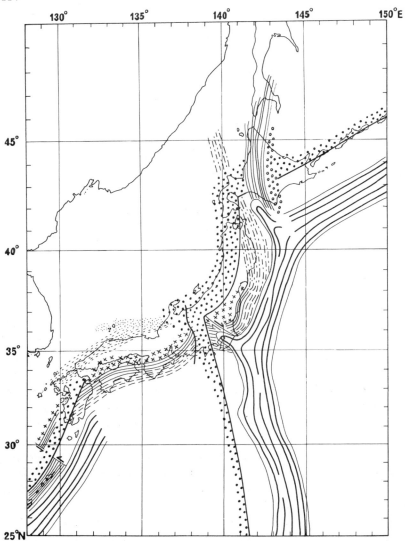

Fig.75. A generalized sketch-map of successive orogenic belts. From younger to older:
{ Closed-circles area = volcanic belt.
{ Heavy-lines area = trench belt (the trench and its land-side subsiding area).
{ Open-circles area = high-temperature metamorphic and plutonic belt of the Alpine-age orogeny.
{ Double-lines area = low-temperature metamorphic belt and its outer (western) depression of the
  Alpine-age orogeny.
{ Crosses area = high-temperature metamorphic and plutonic belt of the Late Mesozoic Era.
{ Light-lines area = low-temperature metamorphic belt of the Late Mesozoic Era (solid lines),
  Simanto geosyncline (dashed lines), and their northern extension (dash-dot lines). (The left-lateral
  fault near the southwestern corner of the map is estimated by Konishi, 1965; and the right-lateral
  fault in central Japan is the Tonegawa Line.)
{ Dotted area = high temperature metamorphic and plutonic belt of the Late Paleozoic Era.
{ Short lines area = low-temperature metamorphic belt of the Late Paleozoic Era.

providing the sites of the volcanic activity of Huzi (Fuji), Yatugatake and other volcanoes.

Yabe (1918), who was one of the most active geologists in Japan in the first two decades of this century, found a well-defined fault line that limited the western border of this "great fissure". The fault is known as the Itoigawa–Sizuoka Tectonic Line (Fig.37) but is often misnamed "Fossa Magna fault". However, no one has been successful in finding the eastern border of Fossa Magna and no definite location has been accepted up to the present. Moreover, the geology of Japan has later been found not to be so simple that it can be divided into northeast and southwest by a "clear fissure" belt. There are many pieces of evidence that argue against this kind of simplicity; the Kanto Range east of Fossa Magna provides the most notable evidence. The Kanto Range (*3a'* and *3b'* in Fig.65) is geologically the continuation of the Outer Zone (ranging from Sanbagawa belt to Simanto belt) of southwest Japan. If Fossa Magna were the central boundary, the Kanto Range, a part of southwest Japan, would be an orphan in northeast Japan.

Recently another view that would overcome this difficulty has been developed, following the steps pioneered by Otuka (1939), who combined the Fossa Magna in the south and the Uetu area in the north into one folded zone (see p.101). Fig.75 shows the new interpretation along this line presented by Sugimura et al. (1963) and by Matsuda et al. (1967). The Mizuho orogeny trends along the Northeast Honsyu Arc and bends southward at central Japan to the Izu–Mariana Arc. Thus the folded belt crosses the Honsyu Island in the middle. The Itoigawa–Sizuoka Tectonic Line is one of the faults running along the western border of the Uetu–Fossa Magna idiogeosyncline in the Mizuho orogenic belt.

Furthermore, the simple concept of the Uetu–Fossa Magna belt is valid for the volcanism since the Early Miocene Epoch up to the present, but it is not valid for the major topographic development. The Uetu depression in the Early–Middle Miocene Epoch extended only to the northern part of the "Fossa Magna" area and ceased its southward extent in the middle of this area. An uplifted part located there is inferred to have divided the "Fossa Magna" into its northern and southern sea basins (e.g., Ikebe, 1957). The southern part behaved as the extension of the Simanto belt, not in volcanic activity, but in the major topographic development, as well as in tectonic movement. Matsuda (1962) named this geological unit "south Fossa Magna".

Thus the concept of "Fossa Magna" seems to have lost the idea of cutting Honsyu Island into two, while "south Fossa Magna" seems to have acquired the meaning of overlapping of the younger Uetu–Fossa Magna belt with the older Simanto belt.

Some remarks on plate tectonics

The boundary between the Eurasia plate and the Philippine Sea plate is estimated on the basis of the present topographical and seismic data. The Nankai Trough off the Pacific coast of southwest Japan is the northeastern extension of the Ryukyu Trench (Fig.76).

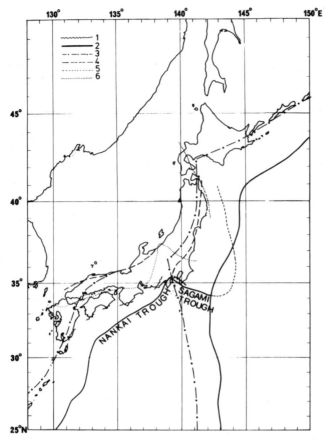

Fig.76. Volcanic fronts and axes of the trenches, both present and past. *1* = Early Miocene Takatiho folding; *2* = axes of trenches, troughs and their linking depression; *3* = fronts of Quaternary volcanic belts; *4* = fronts of Early Miocene volcanic belts; *5* = inferred extension of the mesotectonic trench; and *6* = front of Late Cretaceous and Paleogene acidic igneous area (except earlier Ryoke granitic rocks).

Its deepest part is about 5,000 m and it could not be called a trench because of its relatively small depths. But, it seems to be a former trench filled by later sediments, which are detected to be 1,500 m thick beneath the bottom of the trough (see p.16). Towards its northeastern end the trough becomes shallower and the trace of the extension appears to land to the north, probably owing to the uplift of the Izu–Ogasawara island arc belt during the Quaternary Period (see p.138).

On the other side of the crest of the Izu–Ogasawara Arc, the Sagami Trough extends trending east-southeast and joining with the Japan Trench at the eastern end. There are many who regard the Sagami Trough as a branch of the Japan Trench because of this junction. But, as discussed on p.107 and p.140, the authors have the view that the branch of the Japan Trench depression area should rather extend to the Kanto Plain north of the Sagami Trough, on the basis of the distribution of the negative gravity anomalies. The Sagami Trough has many of the same features as those of the Nankai Trough. Among them the following two appear to be important: the parallel trending to the Japanese main orogeny and the major earthquake zone associated with a similar type of crustal movement. The parallel features are shown in Fig.76. Line 6 in this figure is the outer limit of the Late Cretaceous and Paleogene granitic rocks except Ryoke granitic rocks of earlier age, and the short lines 1 indicate the places where the Early Miocene Takatiho folding is found. The parallelism of these features with the Nankai Trough has often been suggested, but the parallelism with the Sagami Trough has not been put stress upon yet. The seismic epicenter area with shocks of magnitude of 8.0 or larger, associated with prominent earth movement along the Pacific coast of southwest Japan, extends in the belt between the coast and the Nankai Trough and roughly parallels to the mesotectonic trend. This area appears to include the Sagami Trough. It may thus be justified that the Sagami Trough is the eastern extension of the Nankai Trough and is the fossil trench as well. The Sagami Trough is not expected to have resulted from the sinking simultaneous with that of the Japan Trench.

The belt connecting the northeastern end of the Nankai Trough with the northwestern end of the Sagami Trough curves with the convex side to the north and passes through the northern neck of the Izu Peninsula. There is a thick pile of marine sediments deposited at a depth of a few hundred metres in the Late Miocene through the Early Pliocene. The Early Miocene sediments of the trench floor may be expected beneath this formation.

The Ryukyu Trench, the Nankai Trough, and the Sagami Trough are suggested as marking the boundary between the Eurasia plate and the Philippine Sea plate, whereas the Kurile, Japan, and Izu–Ogasawara trenches mark the western boundary of the Pacific plate. Along the Ryukyu Trench, the downgoing of the Philippine Sea plate is continuing even at present. Along the Nankai Trough, the sinking of this plate had a sufficient rate in the Early Miocene time when the Takatiho folding (mark 1 in Fig.76) and the volcanic activity in the northern zone of Southwest Honsyu Arc (the front of the volcanic belt is indicated by line 4 in Fig.76) were in action. The sinking, however, has become slower in rate and has come to contain a right-lateral component in recent years. The slower rate can be inferred from the fact that there are no active volcanoes along the Southwest Honsyu Arc, except those located in the eastern part. The latter volcanoes are believed to have originated from the sinking of the Pacific plate (Fig.38). The right-lateral component can be clearly expected by two pieces of evidence: the right-lateral character of the Median Tectonic Line during the Quaternary (see p.136), and the right-hand faulting estimated at the 1923 Kanto earthquake focus (Imamura and Kishinouye, 1928). The relative movement between the Eurasian and Philippine Sea plates at present thus will cause a right-lateral slip at Sagami Trough also. Mogi (1970) interprets the occurrence of large earthquakes in this area as being due to compressive forces exerted by the underthrusting slab. In this model too, the Nankai Trough is the place where underthrusting is taking place. Mechanism investigation of large earthquakes in this area is highly desired. One solution by Katsumata and Sykes (1969) for a shock off the Kii Peninsula showed a thrust oblique to the Nankai Trough. The direction of the thrust was consistent with the model which shows that the Philippine Sea plate is moving in a northwesterly direction.

The second remark is on the evidence by which the sinking movements of both sides of the Japan Trench may be estimated. The immediate boundary between the Eurasia plate and the Pacific plate is, without doubt, along the Japan Trench and its northeastern extension. This appears to have been effective since the earliest Miocene and the boundary seems to have formed in a different way from that during the time prior to the Miocene Epoch. Line 5 in Fig.76 shows the possible original position of the boundary during the mesotectonic age according to the authors' view. This curve was drawn on the assumption that the mesotectonic foredeep for northeast Japan was the extension of the Nankai and Sagami troughs and parallel to the general trend of the main orogenic belt. The lithosphere of the inner side of line 5 is supposed to be continental, and it is here suggested that the part out of the present trench axis has been consumed by the astheno-sphere underneath the landward slope of the trench since the earliest Miocene by the eastward drift of the northeast Japan lithosphere. This idea is favourable with regard to the following aspects:

(1) The space large enough to connect the southwest Japan portion of the main orogenic belt with its northern extension in northeast Japan could be available.

(2) Although the difference in level between the crest of the northeast Honsyu Arc and the bottom of the Japan Trench is almost the same as those for the Kurile Arc and the Izu–Mariana Arc, the absolute levels of both the crest and the bottom are much higher than those for the other arcs (Fig.93, p.139). This may be due to the existence of the light continental crust not only in the Northeast Honsyu Arc but also beneath the Japan Trench.

(3) Whereas the shallow earthquakes do not occur frequently along the Izu–Mariana Arc, they do occur frequently and have large energy off the Pacific coast of northeast Japan (Fig.25). This observation could be interpreted as due to the buckling or the intense bending of the continental lithosphere off northeast Japan.

(4) The buckling concept for the continent side of the trench is based on the idea proposed by Nakamura (1969). He indicated that the continental lithosphere of the Northeast Honsyu Arc might have moved eastward relative to the underlying astheno-sphere where the primary magma is generated, and consequently the Early Miocene volcanic front might have moved from the same position as that of the recent front (line 3 in Fig.76), to the eastern position (line 4 in Fig.76), should the magma generation zone have remained unchanged. From the interval of the two fronts in his map (Fig.77), a displacement of 10 km in the north and 50 km (0.2 cm/year) in the south may be obtained. The difference between the northern part and the southern part seems to harmonize with the configuration of line 5 in Fig.76.

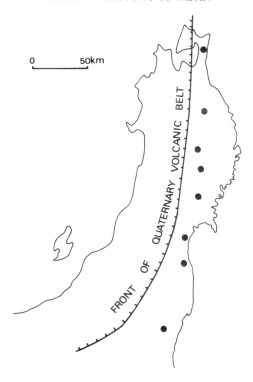

Fig.77. Evidence of the shift of the volcanic front since the Early Miocene along the northeast Honsyu Arc. Solid circle: Early Miocene volcanoes along the Miocene volcanic front. (After Nakamura, 1969.)

The last aspect suggests that the northeast Japan continental lithosphere has undergone an anti-clockwise rotation of about 5° or less since the Early Miocene Epoch. It may be the continuation of the remarkable anti-clockwise rotation of northeast Japan during the mesotectonic age (see p.74), which should have resulted in the major right-lateral faulting along the Tonegawa Tectonic Line (Fig.75). The Tonegawa Line has for more than a few decades been suggested to have been formed during the mesotectonic age, before the "Fossa Magna" formed in the neotectonic age. The latter is not a simple rift zone but has a complicated nature, being a part of the inner idiogeosyncline developed since the Early Miocene Epoch. The formation of the "Fossa Magna" has no essential relation to the anti-clockwise rotation of northeast Japan.

The Japanese main orogeny in the mesotectonic age may be related to the underthrusting of both the Philippine Sea plate and the Pacific plate, which appear to have acted jointly, unlike the present state. In the Middle Tertiary the Philippine Sea plate became independent and this resulted in the drastic change in the geological history of Japan. The independence is regarded as being almost of the same age as the disappearance of the Darwin Rise. This would mean that the Darwin Rise, if it really existed, might have been an important factor for the main orogenic movements in Japan.

The third remark is concerned with the inferred disappearance of a continental plate along the Nankai Trough. Harata (1965) showed, on the basis of the sedimentary structures, that parts of the source of the Paleogene sediments in the Simanto geosyncline were located to the south off the present Pacific coast of the Kii Peninsula. Tokuoka (1967) found a number of orthoquartzite pebbles in the same sediments, which may have derived from a southern land area in the Paleogene Period. The orthoquartzite itself may in turn have been formed in the unknown age and may indicate the source area composed of continental crust. These observations and inferences are concisely summarized and discussed by the Kishu Shimanto Research Group (1968).

The present authors suggested already (see p.110) that the Early Miocene underthrusting of the oceanic lithosphere was so intense that the generation of heat was great enough to produce the volcanic belt in the northern zone (Fig 73) and that the deformation caused the Takatiho folding along the Pacific coast (lines *1* in Fig.76). It is here suggested that this deformation may have been associated with the submergence of the Paleogene land area along the Nankai Trough, which is assigned as the boundary between the Eurasia plate and the Philippine Sea plate.

During the mesotectonic age, more intense underthrusting along almost the same boundary can be inferred from the paired metamorphic belts of Sanbagawa and Ryoke (see p.53). We know that there is a possibility that the continent side lithosphere of the Japan Trench might have buckled down into the asthenosphere at a much slower rate than that of the sinking of the oceanic lithosphere off the Pacific coast of northeast Japan. It might be generally accepted that the underthrusting of the oceanic lithosphere drags the other side block downwards. Just like the absorption of a part of the northeast Japan continental crust into the asthenosphere during the neotectonic age, all parts of the continental crust inferred by Tokuoka and his colleagues might have been absorbed into the deeper part beneath the inner side of the Nankai Trough during the mesotectonic age.

The shallow earthquakes with large magnitudes occur frequently under the continent side sea-floor of the Nankai Trough, in spite of the fact that the Southwest Honsyu Arc is not a typical island arc and that the oceanic lithosphere in the south does not show a rapid sinking as estimated by the submarine topography and other features. The reason for the high seismicity in this area is suggested to be the same as in the continental crust absorption area along the Japan Trench. We have no such high seismicity zone of shallow shocks with large magnitudes along the Izu–Mariana Arc (Fig.25).

Neogene tectonic maps

Many of the Tertiary strata in Japan are folded. Otuka (1933) calculated the contraction coefficient $(F-L)/F$ for the Tertiary strata in Japan and its environs, where F is the length of a curve indicating a folded bedding plane on a geological cross-section showing the length L. He found seven well-defined intensely folded zones of high contraction coefficients as follows (Otuka, 1937):

(*1*) Karahuto–Yezo folded zone. (Miocene–Pliocene; *2a* in Fig.65).

(2) Mizuho—Fossa Magna folded zone (Uetu—Fossa Magna zone in this monograph). (Pliocene—Pleistocene; *1c* in Fig.65.)

(3) Ooigawa folded zone. (Early Miocene; east half of *3a* in Fig.65.)

(4) Sinzi folded zone. (Oligocene—Miocene; Fig.63.)

(5) Amakusa folded zone. (Cretaceous—Eocene; west half of *3b* in Fig.65.)

(6) Ryukyu folded zone. (Oligocene—Miocene; west half of *3a* in Fig.65.)

(7) Taiwan folded zone. (Oligocene—Miocene and Pliocene—Pleistocene.)

The ages in the parentheses are of the most active folding. Otuka (1939) worked out the biostratigraphical division of the Tertiary sequences into successive stages, and illustrated the shoreline changes in Japan and its environs during these ages. He also summarized "Tertiary" (later it was found to include Quaternary) crustal deformations in these areas. It was revealed that the youngest folding took place in the Uetu—Fossa Magna zone and in the Ryukyu—Taiwan zone. The Plio-Pleistocene folding in the Uetu—Fossa Magna zone is called Mizuho folding (Ikebe, 1956).

Such a distribution of folding in space and time is in agreement with our view concerning the sequence of orogenic belts described in preceding chapters. The Uetu—Fossa Magna folded zone represents the continent side geosyncline of the Mizuho orogenic belt (active since the Miocene), the ocean side geosyncline of which extends along the Japan Trench. The Ryukyu—Taiwan folded zone seems to be similar to the Uetu—Fossa Magna zone and related to the deepening of the Ryukyu Trench. The above two zones are certainly formed as parts of island arc structures. The Karahuto (Sakhalin)—Yezo folded zone (*2a* in Fig.65) extending north—south in central Hokkaido appears to have been the small geosyncline that developed in front of the rising Hidaka Range during the Paleogene (Iijima, 1964). Later in the Miocene and Pliocene epochs the sediments were so intensely deformed that in places Decken structures were formed (Otatume, 1941). The zone represents the west side geosyncline of the Hidaka orogeny, which seems to follow the Alpine tectonic cycle (Hunahashi, 1957). The Ooigawa zone and the Sinzi zone are situated on both sides of Southwest Honsyu Arc and show the Oligocene—Miocene folding which was probably the continuation of the main orogeny of Japan. The Ooigawa zone is a part of the Takatiho folding belt.

Matsuda et al. (1967) devised a new index of degree of deformation, which is calculated from a geologic section as in the case of the contraction coefficient of Otuka. In a geologic section, a curve of a selected bedding plane is marked, and then the degree of deformation is defined as $\Sigma|\Delta H|/L$, where ΔH is the amplitude of folding for one wing of folded structure and the throw of a fault, and L is the length of the section. Fig.78 shows the contours of the degree of deformation since the Miocene, based on geological sections published mostly in the scale of 1:50,000. Two highly deformed zones are distinctive in this map. The northern one corresponds to the Karahuto—Yezo zone of Otuka. The folded structure in this zone was produced during the late stage of the Hidaka orogeny. The other highly deformed zone represents the Uetu—Fossa Magna idiogeosyncline repeatedly referred to. For the other three folded zones of Otuka, our degree of defor-

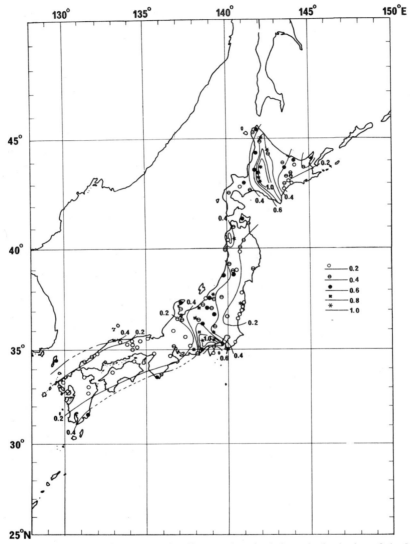

Fig.78. Degree of deformation of sediments deposited since the beginning of the Miocene. (After Matsuda et al., 1967.)

mation is not so high. This is a natural consequence, because the degree of deformation was given for the strata deposited since the Miocene and the foldings of the other zones are considered to have ceased before the Middle Miocene. Putting aside the "Peri-Hidaka" Karahuto—Yezo zone, the zone of intense folding is restricted to the inner side of the east Japan island arcs (Fig.78).

Fig.79 shows the general trends of the Late Cenozoic fold axes as well as the areas

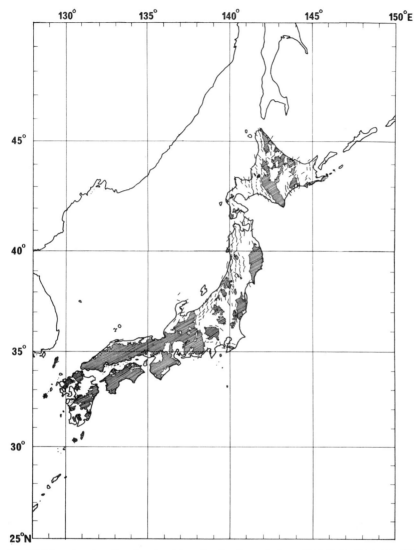

Fig.79. Map showing the structural trends of sediments deposited since the beginning of the Miocene and the exposure of their basement rocks (hatched areas). (After Matsuda et al., 1967.)

where older rocks are exposed. Each short line in Fig.79 indicates the general trend in quadrangular areas. It is given by the average azimuth of the fold axes and the fault lines mapped on a geologic sheet-map which covers about 24 × 29 km² in area. In general, these azimuths of fold axes are parallel to the trends of island arc topography and are not parallel to those of the structures of older rocks, except in two of the "crossing districts", i.e., central Hokkaido and south Fossa Magna, where the Late Cenozoic fold axes are parallel to the trends of the older belts.

Fig.80 shows the amount of vertical movement since the Miocene up to the present.

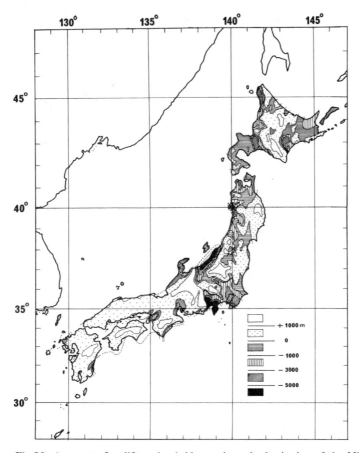

Fig.80. Amount of uplift and subsidence since the beginning of the Miocene. (After Matsuda et al., 1967.)

Here it is assumed that the amount equals the present height of the unconformity between marine Miocene strata and older rocks. Extrapolations in the areas where the Miocene strata were neither deposited nor left against erosion are carried out from the neighbouring data and the levels of younger erosion surfaces. In this map of vertical movement, the distinct belt of depression is also recognized along the continental half of the east Japan island arcs.

QUATERNARY TECTONICS IN JAPAN

"Plio-Pleistocene movements"

Prior to a revolutionary change in the idea of the age of the tectonic movements based on the progress of Quaternary tectonics in the last decade, the "Plio-Pleistocene movements" were believed to be the last conspicuous tectonic movements in Japan.

For example, the folding in the Uetu zone, the northern and middle part of the Uetu—Fossa Magna idiogeosyncline, was assigned to be of Plio-Pleistocene age by Otuka (1937) as mentioned on p.121. In this zone, most of the folds involve uppermost Neogene beds and in places involve the Plio-Pleistocene formation. Many of the hills extending parallel to the general trend of the fold axes represent the anticlinal cores of folds or the uplifted fault blocks. The longitudinal valleys and basins, some of which are now alluvial plains, are also found at the site of the synclinal axes or the subsiding blocks. These observations support the idea that the folding of this zone is so young that the topography corresponds to the geologic structure. The Late Pleistocene terrace surfaces are also deformed and consequently the folding has been regarded as active (Otuka, 1941). Because of the much smaller amount of deformation of the terrace surfaces than that of the Neogene strata, this folding was concluded to have occurred in the Plio-Pleistocene age and the recent deformation was regarded only as an after-effect of this major folding or at most as occupying a small portion of the whole folding. The folding in the Uetu zone was named the Mizuho folding.

The second example is provided by the faulting in the Kinki district (Fig.63) named the Rokko movement (Huzita, 1962; Ikebe and Huzita, 1966). The faults remarkably displace the Plio-Pleistocene formation in this district. On the other hand the displacements of the Late Pleistocene terrace surfaces by the faults are only rarely found, and are very small in comparison to those in the Plio-Pleistocene formation in this district. Thus, the peak of the Rokko movement was concluded to be of the Plio-Pleistocene age.

The Mizuho folding as well as the Rokko movement was believed to be mainly of the Plio-Pleistocene age. However, this view will be shown to be short-sighted when the results of Quaternary tectonics are comprehended in the following sections.

Quaternary tectonic maps

It has been attempted since 1963 by A. Sugimura and his colleagues to summarize the basic geological and geomorphological information available for the study of the Quaternary earth movement in Japan, and to illustrate it by Quaternary tectonic maps, which will here be referred to, and which will show the distribution of the Quaternary warping rate and active faults and folds in Japan. The maps have been published in a series of five papers. The first paper by Hatori et al. (1964) includes, beside preliminary maps, a list of the main sources and some discussion of the Quaternary earth movements. The second paper by Kaizuka et al. (1966) is a progress report which gives, beside the maps, geomorphological and geological cross-sections through four representative locations of uplift, subsidence, faulting and folding. The third paper (Research Group for Quaternary Tectonic Map, 1968) is a summary of their work, describing the method of preparing the maps and showing five maps in a concise form. Fig.81—83 are three of the maps. The fourth paper (National Research Center for Disaster Prevention, 1969)

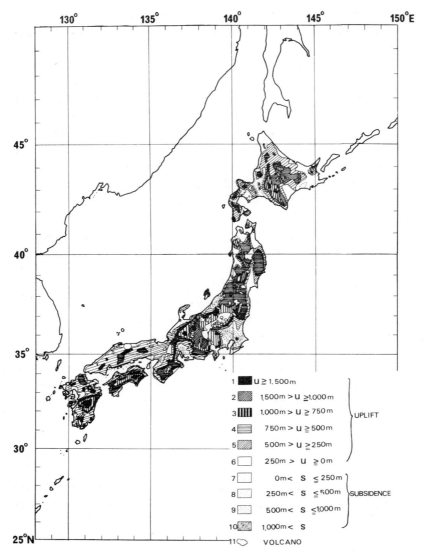

Fig.81. Amount of uplift and subsidence during the Quaternary period. u = uplift, s = subsidence. (Research Group for Quaternary Tectonic Maps, 1968.)

includes six maps with a scale of 1:2,000,000. Three of these are contour maps showing amounts of Quaternary uplift and subsidence, the next two show the distribution of Quaternary faults and folds, and the last one is the "gipfelflur" map of Japan. The fifth paper (National Research Center for Disaster Prevention, in preparation) provides the table of the detailed data of uplift, subsidence, faults, and folds throughout Japan.

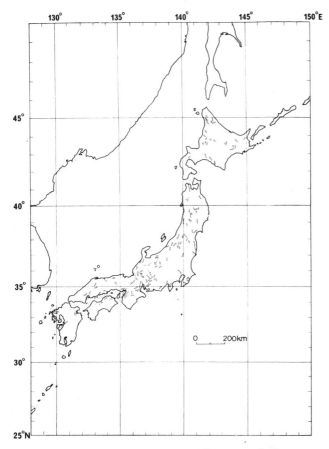

Fig.82. Map showing the distribution of Quaternary faults.
Line with teeth = dip-slip fault, teeth-side being the down-thrown side; line with an arrow = strike-slip fault; and hatched area = group of parallel faults. (Research Group for Quaternary Tectonic Maps, 1968.)

The amounts of uplift and subsidence during the Quaternary Period are estimated on two criteria: one being geomorphological and another geological. The team of geomorphologists has plotted the present heights of the erosion flat surfaces formed during the latest Tertiary or the earliest Quaternary age. It is assumed that the surfaces were formed not far from sea-level and have been kept largely unchanged by denudation or deposition. The team of geologists has plotted the present altitudes and the depths of the locations where Upper Pliocene to Lower Pleistocene marine sedimentary formations outcrop or are found by means of drilling to be buried. It is assumed also in this case that the sedimentation took place at shallow depths. The vertical displacement thus estimated represents approximately the algebraic total since the beginning of the Quaternary up to

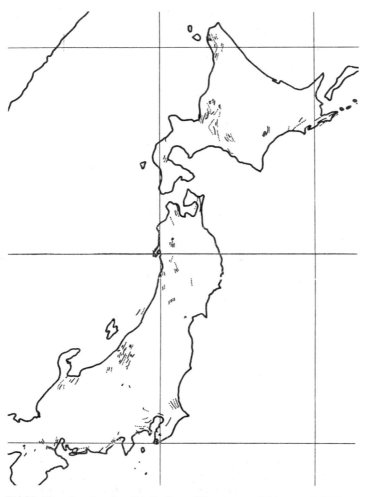

Fig.83. Map showing the distribution of Quaternary fold axes. Solid line = anticlinal axis; and dashed line = synclinal axis. (Research Group for Quaternary Tectonic Maps, 1968.)

the present. The errors derived from these assumptions seem to be smaller than those derived from the roughness of time estimations, which dates the Plio-Pleistocene Age from 3 m.y. ago to 1 m.y. ago. Taking advantage of the two methods, geomorphological and geological, Fig.81 was synthesized.

Both the faults distribution map (Fig.82) and the folds distribution map (Fig.83) are drawn on those faultings and foldings that have deformed sedimentary beds and terrace surfaces since Upper Pliocene age. Faults more than 1 km in length and folds 0.5 to 30 km in wave-length are selected. The regional trend of fold-axes and strike of faults shows a characteristic pattern in harmony with the island-arc features.

Generally speaking, mountain districts have been uplifted, while lowlands have subsided. The maximum uplift of about 1,700 m during the Quaternary is found in the Central Ranges (Japan Alps), and the maximum subsidence of about 1,400 m in the Kanto Plain, both in central Japan (Fig.81). These amounts are distinctly large for the interval of 2 or 3 m.y., which is a small fraction of the whole Late Cenozoic. It is concluded that some changes in the rate of deformation must have occurred in the Pliocene or Pleistocene and that the present relief of the Japanese Islands is largely due to the Quaternary tectonic movements. The birth and development of the earth movements characteristic to the Quaternary will be discussed in the following sections.

Three examples characterizing the Quaternary tectonics

Three examples dealt with here are the Kanto subsidence, the Oguni folding (a local part of the Mizuho folding), and the Atera faulting. All these examples show geological features deformed subsequent to the Late Pliocene as well as geomorphological features deformed in the Pleistocene and Holocene time.

The rate of the Kanto basin-forming movement was discussed by A. Sugimura and his colleagues (Kaizuka et al., 1963, p.126; Naruse and Sugimura, 1965, p.363; Naruse, 1968). We have a Riss—Würm or Sangamon Interglacial marine terrace surface widely distributed in the Kanto area. A circular basin-forming earth movement is clearly shown by a contour map of the terrace surface (Fig.84). The age of this terrace is figured to be about 125,000 years (Machida and Suzuki, 1971). We have additional data on the configuration of a certain horizon in the marine Plio-Pleistocene Kazusa Group. The horizon used is a pyroclastic key bed (named U6) that is believed to be at the approximate boundary between Pliocene and Pleistocene, as indicated by the first predominance of cold-water planktonic Foraminifera in the latest Cenozoic section. Kawai (1965, 1967) prepared an isobath map of the top of the Umegase Formation, which is a horizon about 200 m above the U6 bed. If we assume that the U6 bed was deposited on a 200-m deep sea-floor and the formation was initially horizontal, the isobaths would approximately indicate the amount of subsidence since the beginning of the Pleistocene Epoch. A circular basin-forming movement is also demonstrated on this map of isobaths. Thus, there is an assemblage of data concerning the rate of tilting of the ground in a 35-km long zone between Tiba (Chiba) and Kururi in southern Kanto during different time spans. These data are summarized in Table VII. Tiba seems to be situated near the center of downwarping and Kururi near the boundary line between a subsided area and an uplifted area. Therefore, the differences in vertical displacement in Table VII between these two localities would be close to, but may be a little larger than, the maximum amount of subsidence in Kanto. The estimated ages may contain some errors, but at least the order of the rate of ground tilting during the three different time spans from 6,200 years to $2 \cdot 10^6$ years seems to be the same. It follows that the downwarping can be regarded as steady during at least 10^5 years. The average rate of tilting is about 0.1 min/1,000 years. An east—west geologic section of the Kanto Plain was prepared by Naruse (Kaizuka et al.,

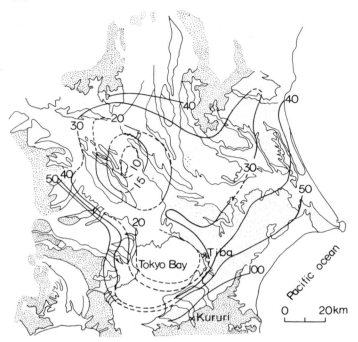

Fig.84. Map of the Kanto Plain, showing the amount of uplift in metres of the marine Simosueyosi terrace surface, relative to the present sea-level. Dotted area = marine Simosueyosi Terrace; and hatched area = land area in the Simosueyosi age. (After Naruse, 1966.)

TABLE VII

Rate of warping in the Kanto area

Time span (years)	Tiba	Kururi	Difference	Rate
6,200	+4 m	+11 m	7 m	1.1 m/1,000 year[1]
125,000	+20 m	+140 m	120 m	1.0 m/1,000 year[2]
2,000,000	−1,200 m	−200 m	1,000 m	0.5 m/1,000 year[3]

[1] Sitamati terrace. Data from Sugimura (1962a, fig. 1, 2). The height at Kururi is obtained by extending the Sitamati terrace level from the coastal areas.
[2] Simosueyosi terrace. Data from Naruse (1966).
[3] Top of the Umegase Formation or lower part of the Kazusa group. Data from Kawai (1965).

1966, fig.3). According to this section, the base of the structural basin filled by the sedimentary strata of the Kazusa Group is about twice as deep as the downwarped key bed U6. If a downwarping of an almost constant rate is assumed to have formed this structural basin, then the beginning of the downwarping must have occurred twice as many years ago as did the first predominance of cold-water Foraminifera, which is paleo-magnetically confirmed to be about $2 \cdot 10^6$ years old (Nakagawa et al., 1969).

The second example is provided from a folding with north—south trending axes in the

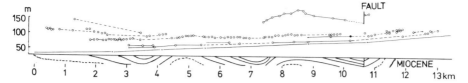

Fig.85. Profile of the Oguni River terraces and geologic section of the Pliocene strata, in the Mogami district of northeast Japan. The profile is ten times vertically exaggerated. Lines in the upper profile = terrace surfaces and a river bed, solid lines indicating good continuation; open circles = altitude measured points; and lines in the lower section = stratification in the Pliocene formation, solid lines representing key beds. Note the deformation of the older terrace surfaces has progressed further. (After Sugimura, 1952.)

Mogami district, Yamagata prefecture, northeast Japan (Sugimura, 1952). The district belongs to the Uetu–Fossa Magna folded zone (Fig.65). The Oguni River, a tributary of the Mogami River, crosses the folded structure from east to west, the structure having the western flanks of the anticline dipping steeply with faults and the eastern flanks dipping gently. Along the Oguni River extend several steps of river terraces. The profiles of the terraces undulate with analogous shapes but with less amplitudes than those of the underlying folds of Pliocene strata (Fig.85). Moreover, the amplitude increases from younger terraces to older ones. This progressive folding behaviour of the river terraces indicates that the folded structure has been growing more or less steadily without having had significant retrogression or stagnation. An age of 35,000 years was obtained for one of the terrace deposits by the radiocarbon method (Yamagata et al., 1973). Under the assumption of a uniform rate of folding, which seems to be acceptable as a first approximation in view of the above-mentioned progressive increase in the degree of deformation, the order of 10^6 years is obtained for the duration period of this folding.

The Atera faulting (see Fig.90, p.137) has a left-lateral component and has displaced the river courses horizontally with offsets of 7–10 km (Sugimura and Matsuda, 1965, fig.1). The topography of both sides of the Atera fault scarp is relatively flat (Sugimura and Matsuda, 1965, fig.2). This flat topography is thought to be derived from an erosion surface, the age of which is supposed to be older than the Plio-Pleistocene Seto Group. Later the surface has been offset vertically about 800 m by the Atera faulting. The Atera fault cuts several terrace surfaces of the Kiso River at Sakasita. The description of the fault trace across the terrace has been fully given elsewhere (Sugimura and Matsuda, 1965). Fig.86 shows that the amount of each horizontal and vertical offset is progressively larger for the older terraces than for the younger ones. This relationship also indicates that repetitive movements with increments of a few metres took place over a long period (Fig. 87). The second oldest terrace shown as *VIII* in Fig.86 is dated by radiocarbon methods to be 27,000 years B.P. (Quaternary Research Group of the Kiso Valley and Kigoshi, 1964). Assuming a steady movement for the Atera fault, the rate of the net slip would be about 5 m/1,000 year on the basis of the amount of offset and radiocarbon age of terrace *VIII*. If the faulting with this rate is assumed to have formed the Atera fault scarp and to have offset the river courses across the fault, it also must have begun about 10^6 years ago.

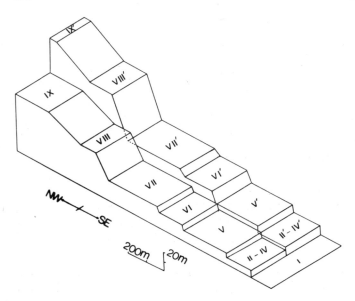

Fig. 86. Diagram showing that the Atera fault cuts seven steps of river terraces at Sakasita. (After Sugimura, 1967a.) Roman numerals indicate the numbers of the terraces. The vertical scale is exaggerated five times.

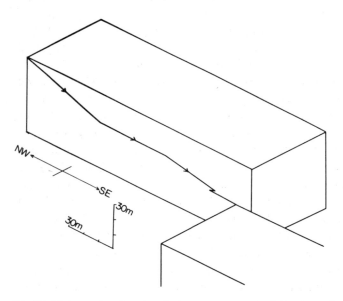

Fig. 87. Diagram showing the displacement vectors of the Atera fault. (Revised from Sugimura and Matsuda, 1965.)

Duration period of Quaternary tectonic movements and volcanic activity

The above stated three examples of Quaternary earth movements (warping, folding and faulting) show the common characteristic that the rate of movement has been almost uniform during the Late Quaternary. Similar examples from other areas also yield evidence of the Late Quaternary deformations with almost steady rates. Yoshikawa et al. (1964, 1968) got the rate of a steady tilting in the southeastern Sikoku district, using the altitude distribution of coastal terraces as well as geodetic data. We have an increasing number of documentations dealing with the time uniformity of earth movements in Japan (Nakamura et al., 1964; Kaizuka, 1967; Ota, 1968, 1969).

If we assume the rates of deformation that vary from place to place but are uniform in the course of time, a remarkable coincidence in the data of the beginning of these movements can be obtained as shown in Fig.88. The time and amount of displacement are given in a logarithmic scale by the abscissa and ordinate in this figure. Broken lines drawn on this diagram indicate: (*1*) the uniformity of the rate of movements; and (*2*) the relative magnitude of the rate. If we take the amount of displacement as $x = x(t)$, the uniformity of the rate, $\frac{dx}{dt}$ = constant, implies:

$$\frac{d(\log x)}{d(\log t)} = \frac{t}{x} \cdot \frac{dx}{dt} = 1 \tag{7}$$

Therefore the time uniformity is indicated by the angle of $45°$ between the broken lines and the axis of coordinates and this relation is independent of the rate. The amount of the rate is known by the position of the line. As the rate increases, the line becomes increasingly close to the upper left corner. The position of the lines in Fig.88 is fixed by the data points plotted with circles and crosses, in which the values of x are known from terrace deformation and t from radiocarbon dates and the estimated age of $9 \cdot 10^4$ years for the Simosueyosi Age which should be $12.5 \cdot 10^4$ years according to the recent measurements. The horizontal lines in Fig.88 indicate the total amount of movements on the basis of the evidence cited in the preceding section.

The intersecting points indicated by quadrangles in the diagram would determine either: (*1*) the date of formation of the sedimentary or geomorphological features; or (*2*) the date of the beginning of these movements, both under the assumption of uniform rate. In Oguni, the age of the strata deposited is known to be Pliocene and would certainly be older than the age indicated by the quadrangles. In Atera, the age of the topographic features formed seems to be older than the Plio-Pleistocene Seto Group and would thus be older than the age indicated by the quadrangles. Therefore most of these quadrangles would represent the latter of the above alternatives, i.e., the date when the latest earth movements started. It is certainly remarkable that the five quadrangles in Fig.88 line up in a narrow band around the date 10^6 years ago. This coincidence in age implies that the duration period of the earth movements would be similar to each other in

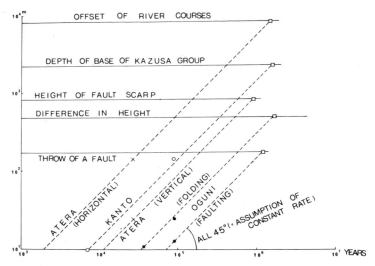

Fig.88. Graph of displacement as related to time both in a logarithmic scale. (After Sugimura, 1967a.) Crosses = Atera fault; open circles = Kanto area; closed circles = Oguni area; and quadrangles = intersecting points.

TABLE VIII

Examples of rate of movement in mm/1000 year

Period (X 10³ year)	Nobi Plain[1]	Kanto Plain[2]	Nobi/1.7; Kanto
6		1,100	1,100
35	1,700		1,000
125		900	900
300 – 600	1,000		600
2,000		500	500
4,000	400		250

[1] After Kuwahara (1968).
[2] Between Tiba and Kururi (Fig. 84); revised from Sugimura (1967a).

spite of the difference in the type of deformation. There is no information about the mode of the beginning of deformation, whether it was abrupt or gradual. But in any case, the order of time necessary for the total amount of movements is not so different from 1–2 m.y.

A second approximation on the problem of the rate of deformation would be possible if more detailed aspects in the change of the rate could be discussed. Here are two examples of the change in the rate of movement, from the Kanto Plain and the Nobi Plain. In Table VIII, toward the lower part, the longer the period concerned is. These examples show an acceleration from the Pliocene and Early Pleistocene time to the Late Pleistocene and Holocene time. Fig.89 has logarithmic scales as Fig.88. If the rate is

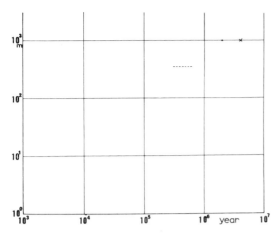

Fig.89. Relation between the amount of subsidence and the time span in the Kanto and Nobi plains of central Japan. Both axes are in logarithmic scale. Solid circles represent data for the Kanto Plain, and crosses and a dashed line are data for the Nobi Plain in which the amounts of subsidence are divided by 1.7 for the sake of accordance with those for the Kanto Plain. Note the rate is uniform since 10^5 years ago but is less for a period of more than 10^6 years. (Data from Kuwahara, 1968, for Nobi Plain and revised data from Sugimura, 1967a, for the Kanto Plain.)

constant, the data points must stand in a line diagonal to both axes by $45°$. But, actually it deviates to the slower position towards the past, i.e., the Pliocene Epoch.

This conclusion supports the comparison between Neogene and Quaternary tectonic movements (Fig.80, 81). Kaizuka and Murata (1969) calculated the difference in amount of uplift and subsidence between the two periods, one since the beginning of the Miocene and another during the Quaternary Period, and concluded that the rate of Neogene movement is far less than that of the Quaternary one. They calculated also the difference in the degree of folding, and reached the same conclusion as in the case of uplift and subsidence. Chinzei (1966, 1968) also states that the latest Pliocene and Quaternary movements in Japan are significantly greater than that of other Late Cenozoic epochs. The tectonic movements must have been activated in the Late Pliocene or in the Early Pleistocene.

It remains to be emphasized that, although we know little about the interrelation between the different types of the tectonic movements, the uplift of most ranges and the subsidence of most basins in east Japan as well as faulting and folding in the Uetu– Fossa Magna geosyncline seems to have taken place during the same period – from the Early Quaternary Period to the present. At least in central and eastern Japan, the rate of every tectonic movements appear to have been accelerated from the Neogene Period towards the Quaternary Period, irrespective of their type.

On the other hand, according to the quantitative study on volume of volcanoes described on p.55, the amount of volcanic products in the Quaternary is larger than that in the Late Neogene (Fig.67). It will now appear probable that the tectonic movement and the volcanic activity alter their intensities roughly parallel to each other. It does not mean, of course, that each movement relates directly to each activity, but it means that the source of energy of both activities in the whole island arc system seems to have become stronger from the Pliocene to Pleistocene. It is tempting to speculate that some important change took place in the plate subduction in the last 1—2 m.y. to cause the acceleration of both the crustal movements and the generation of basaltic magmas, in accordance with the view to be discussed in Chapter 3.

It is interesting to note that the ratios of volcanic products with $1,000 \text{ km}^3/\text{m.y.}$ for the Middle and Late Neogene and with $2,500 \text{ km}^3/\text{m.y.}$ for the Quaternary are in harmony with the comparison between the slip rate of the underthrusting with 9 cm/year for the average Cenozoic Era (Le Pichon, 1968) and that of 22 cm/year derived from the recent seismicity (Davies and Brune, 1971). As illustrated by Matsuda et al. (1967, fig.4), the Late Quaternary Period is one of the two periods of intense activity of tectonism and volcanism in the Mizuho orogeny, the other being the Early Miocene Epoch. Whatever the spreading may be, the underthrusting and therefore the earth movements in island arcs seem to be essentially episodic.

Strike-slip fault system

The distribution pattern and sense of movement of surface faults that formed during historical earthquakes and the Late Quaternary Period in central Japan suggest a stress system producing conjugate sets of strike-slip faults (the right half of Fig.90). Among these faults, the Atera fault (No.V in Fig.90) shows the highest rate of faulting. Recently numerous active strike-slip faults have been found in western Japan (the left half of Fig.90). Among these the right hand nature of the Median Tectonic Line is remarkable (Okada, 1968, 1970). It has the greatest amount of slip of the right-lateral faults found throughout western and central Japan.

Fig.91 shows vectors of faultings for two faults: the Atera fault and Median Tectonic Line fault. It is shown that strike-slip predominates to dip-slip in both cases and that the order of rate of net slip of the two faults is equal to each other. These results suggest that a homogeneous horizontal deformation has occurred in central and western Japan. Because of the almost equal slip in left-lateral faultings and right-lateral faultings, the deformation as a whole would not be the shearing strain but either the east—west contracting strain or the north—south dilatational strain. For central Japan, a tendency of the earth's surface to contract over the past decades is calculated on the basis of results of re-triangulation (Kasahara and Sugimura, 1964a, b), while a tendency to uplift is deduced from the results of releveling (see p.46). Horst-and-graben structures which characterize the Japan Alps and their environs in central Japan seem to be the secondary fault zones that play a role in adjusting the horizontal movement to the regional contracting strain system. Thus, the contracting strain system is more favourable for these features than the dilatational system. Summarizing these strike-slip faults as well as

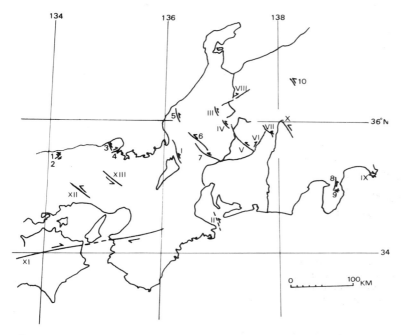

Fig.90. Quaternary strike-slip faults in central Japan. Arabic numerals *1–10* represent 9-2, 9-1, 7-1, 7-2, 11-1, 1-2, 1-1, 8-1, 8-2, and 14-1 in Table III, respectively. *I* = Yanagase fault, *II* = Isewan fault; *III* = Sirakawa fault; *IV* = Mumai-Mugisima fault; *V* = Atera fault; *VI* = Kakizoretoge-Narai fault; *VII* = Kamiya fault; *VIII* = Atotugawa fault; *IX* = Minamisitaura fault; *X* = Sioziri-Nirazaki fault (a part of the Itoigawa–Sizuoka fault); *XI* = the Median Tectonic Line fault; *XII* = Yamasaki fault; and *XIII* = Mitoke fault. (After Sugimura and Matsuda, 1965; and Huzita, 1969.)

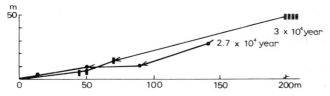

Fig.91. Displacement vectors of the Median Tectonic Line (quadrangles) and the Atera (circles) faults. (After Matsuda, 1969.)

recently uplifted ranges, Huzita (1969) prepared a map of compression trajectories (Fig.92).

According to the calculation by Le Pichon (1968), an east–west shortening with the rate of 9 cm/year is occurring in the environs of Japan across the boundary of the Eurasian plate and the Pacific plate. It is noteworthy that the direction of this shortening is the same as that of the general trend of the contracting axes in central and western Japan. The most part of the shortening would be carried out by the sinking of the oceanic lithosphere into the mantle underneath the east Japan island arc system. Furthermore, the continental lithosphere would be affected by the sinking and would be more or less

Fig.92. Quaternary deformation in central and western Japan. Heavy dashed line = the Median Tectonic Line parallel to the coasts and Itoigawa–Sizuoka fault perpendicular to the coasts; heavy solid line without arrow = axis of an uplifted range; solid line with arrow = strike-slip fault; and fine dashed line = axis of maximum shortening. (After Huzita, 1969.)

compressed. The contraction of the continental lithosphere in central and western Japan is suggested as sharing a part of the shortening across the boundary between the Eurasian plate and the Pacific plate.

Another possibility is that this contraction along the Southwest Honsyu Arc represents the strain exerted in the continental lithosphere by the reducing Philippine Sea floor, which lies between the Ryukyu Trench in the west and the Izu–Ogasawara Trench in the east.

The island arc trench system and Quaternary warping

The similarity in profiles across the island arc and trench between different positions along the Kurile, Japan, and Izu–Ogasawara trenches is well illustrated by Otuka (1938). The warping that has formed the major topography can be approximated to a *mega-folding* for this reason.

The upper curve of Fig.93 is the longitudinal profile of topography along the axis of the east Japan island arcs (Fig.10); the lower curve is that along the axis of the Japan and Izu–Ogasawara trenches. It may be said that there is a parallelism in the lower harmonics of the two profiles and the average difference of heights between the crest of the island arcs and the bottom of the trench is about 6,000–7,000 m. Murray (1945) pointed out a similar parallelism for the Aleutian island arc and the Aleutian Trench. Similar profiles along the whole "Honsyu Arc" (Fig.94) show that the parallelism is interrupted in central Honsyu. These observations seem to justify the view that the "Honsyu Arc" should be divided into two parts, each of which belongs to a different arc–trench system (see p.15). The uniformity of the difference of heights within each system should be noted.

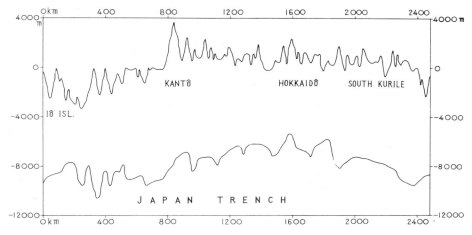

Fig.93. Longitudinal profiles of the topography along the east Japan arcs: upper curve along the axis of the arcs and lower curve along the axis of the trenches.

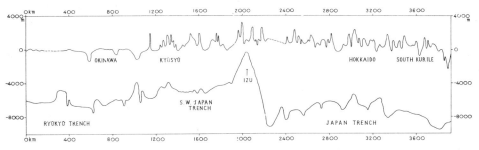

Fig.94. Longitudinal profile of the topography along the Honsyu "Arc": upper curve along the axis of the arcs and lower curve along the axis of the trenches.

On this basis, it is proposed that a mega-folding movement, or a fairly uniform deformation, similar to a unit of an anticline and a syncline, has taken place along the whole length of the arcs and trenches. Upheaval of the central mountains in Honsyu amounting to 1,700 m was observed to have occurred in the last two or three million years (see p.135), whereas the simultaneous subsidence of the Japan Trench by some 4,000 m will be substantiated below. The long wavelength undulation ($\sim 10^3$ km) along the arcs in Fig.93 is probably the relic of topography before the mega-folding movement in question. For instance, the central part of Fig.93, the northeast Honsyu and probably the Japan Trench areas are suggested as having been a part of the older orogenic belt (see Fig.75, 76), and hence to have been originally higher because of the light continental lithosphere as discussed on p.118. The proposed type of the later mega-folding deformation was shown to be produced by a descending convective flow (e.g., Griggs, 1939), although we need a more thorough dynamical investigation on the simultaneous occurrence of the uplift of island belts and subsidence of trench belts by a downward flow along the inclined plane.

It was established in earlier sections that, on the land area the Quaternary movements are more intense than the preceding ones. Is the tectonic activity in the Quaternary on the sea-floor similar to that on land? Almost nothing was known about the sea-floor until the last ten years. For instance, there were no data or idea at all as to the date when the Japan Trench was formed. However, with the dredging of sand and pebbles by the R.V. "Ryofumaru" of the Japan Meteorological Agency during the JEDS (Japanese Expedition of Deep Seas) cruises, information on the history of the trench started becoming available (Iijima and Kagami, 1961). The pebbles collected at the sea-floor 2,300 m deep on the slope down to the Japan Trench off the Pacific coast of northeast Japan contained many fragments of siltstone and sandstone, which were concluded to have been deposited in a shallow sea in the latest Miocene or earliest Pliocene, the dating being based on the diatom remains in the siltstone. The evidence that the fragments are too large to be deep-sea sediments and have holes bored by shallow sea organisms, indicated that these fragments were not transported from the present coastal area, but that the area around the dredge location was the floor of a shallow sea near the coast from the time of deposition of the siltstone to at least the Middle Pliocene. Apparently this area must have subsided to the present depth later. The subsidence of this area would have commenced probably after the end of the Pliocene in accord with the acceleration of tectonic movements observed in the land area (see p.136). This subsidence might have a direct relation to the deepening of the Japan Trench itself. The recent seismic profiler exploration (Fig.95) made it clear that the floor on the island side of the Japan Trench has thick younger sediments and many younger faults cutting through these sediments (Ludwig et al., 1966), showing the very recent tectonic movement presumably caused by the sinking of the oceanic lithosphere. This indicates that the subsidence of the floor may be still in progress.

The negative isostatic gravity anomaly belt, running just inside the Japan Trench, may be related to the recent forced subsidence of the trench. As noted earlier, there are two remarkable bulges of negative anomaly belts on the trench side of the intersections of arcs, that is, at the southern tip of Hokkaido and in the southern Kanto area (Fig.96). Geological evidence indeed shows that these two areas have recently subsided. The southern Kanto area shows a Quaternary subsidence with an almost constant rate of the order of 0.5–1 m/1,000 years (see p.134). This subsidence was activated about 1–3 m.y. ago and is still active at present. Combining these indications concerning the crustal movement of the land and water areas, the authors propose that the subsidence of the Japan Trench was also activated about 2–3 m.y. ago, and that this subsidence is still active at the present time. The total amount of subsidence is estimated to be 3,000–4,000 m near the axis of the trench, on the basis of the depth of the flat top of seamount Sisoev located on the outer slope of the Japan Trench (Uyeda et al., 1964). If we assume that the rate of subsidence has been constant, it would be 1–2 m/1,000 years.

The recent subsidence of the Japan Trench becomes more acceptable when we consider that the uplift of the mountains in East Japan may be the counterpart of the subsidence, and that the greater part of the uplift did occur during the Quaternary. From the study of the Neogene geology of an area in northeast Japan, Chinzei (1966) offers clear evidence that the emergence of the portion of the Oou Range above sea-level began

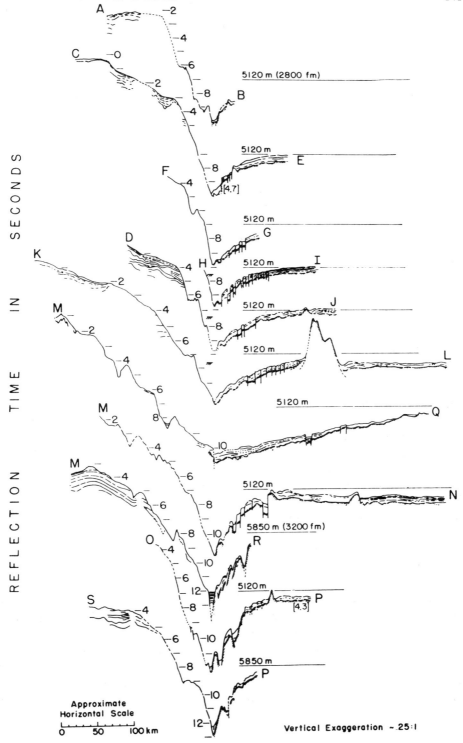

Fig.95. Seismic reflection profiles across the Japan Trench. From above to below, the profiles cover the latitudes from about 41°N to 31°N. Vertical scale represents two-way reflection time in seconds (1 sec. = 400 fathoms or 730 m). (After Ludwig et al., 1966.)

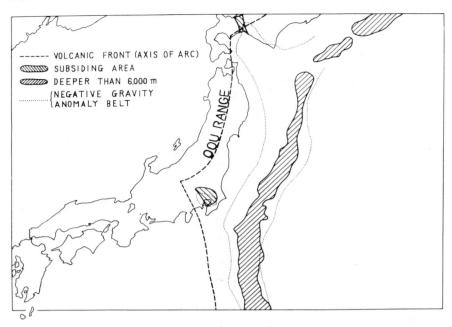

Fig.96. Map of the Northeast Honsyu Arc and the Japan Trench with the corresponding negative gravity anomaly belt. A marked deficiency of mass is revealed beneath the trench belt and two subsiding areas (the Isikari and Kanto plains).

only in the Late Pliocene. He concluded that the main uplift, amounting to 500—1,000 m along the Oou Range, took place after the Late Pliocene. He has also summarized similar pieces of evidence showing that the latest uplift of many ranges from west Japan to the Kitakami-Abukuma geanticline occurred during the Quaternary Period (Chinzei, 1968). The volcanoes in the east Japan volcanic belt are mostly on the crests of this upwarping. The front of the volcanic belt along the Northeast Honsyu Arc coincides with the Oou Range. The relationship between upwarping and volcanic activity was probably close, as shown by the spatial as well as temporal coincidence between them as discussed in the last section.

Most of the Late Cenozoic folded zones described earlier (see the section *Late Cenozoic history of Japan*, pp. 98—124) are still active. This is indicated by folded Quaternary strata and terrace surfaces as exemplified by the Oguni area (p.131). The geographical distribution of the rate of foldings has recently been summarized by Kaizuka (1968). According to him, the rates of movement are as much as 10 times higher in east Japan than those in west Japan, where only a few active folds occur on the Pacific side, mainly in the Simanto belt. It seems to the authors that this difference can be explained by the difference in age of the culmination of folding: in east Japan it is the Quaternary and in the Simanto belt it is the Takatiho phase (Early to Middle Miocene). It is interesting that, as Kaizuka also points out, the folding axes in east Japan are almost parallel to the Japan Trench and those in west Japan to the Nankai Trough. The latter is about 5,000 m deep below the crest of the island arc that parallels it, and the Ryukyu Trench, the southwestern extension of it, shows 6,000 m depth below the crest of the Ryukyu Arc, whereas the Japan Trench has the relative depth of 7,000—8,000 m below the crest of the associated arcs (see Fig.93 and 94). The differences in the rate of folding and the relative depth of trench are in harmony with other characteristic values of island arc activity (Table X, p.189).

The major relief of the Japanese Islands seems to be representing a triple junction of three island arcs: the northeast Honsyu, the southwest Honsyu, and the Izu—Ogasawara arcs. But, it is not the case since they have not been formed by a simultaneous warping. The Proto-Honsyu Arc continental lithosphere was thickened by the main orogeny in the mesotectonic age. The geometry of the plate movement changed suddenly at the beginning of the neotectonic age, when the east Japan island arc system began to operate, including the northeast Honsyu and Izu—Ogasawara arcs. The major relief is strongly controlled by the mega-folding along this system, which is believed to have occurred essentially during the Quaternary Period.

CHAPTER 3

Processes under Island Arcs

We have seen that the island arc system of Japan may be divided into the east Japan arcs and the west Japan arcs and that the former system has the more typical island arc activities. In Fig.97, an idealized cross-section of the east Japan arcs is shown, summarizing the various zonal features.

In this chapter an attempt will be made to explain the island arc features described in Chapters 1 and 2 by the hypothesis of a descending mantle flow. The proposition of this flow of the upper mantle by the authors was made before and is therefore independent of the present popularity of the sea-floor spreading or plate-tectonic theories, but it may well be a part of the flow system that is assumed to be the cause of the spreading of ocean floors and motions of plates.

CONVECTION CURRENT DESCENDING UNDER ISLAND ARCS

Nature of the convection current and viscosity of the mantle

As far as Newtonian viscosity is assumed for the mantle material, Bénard-Rayleigh type convection is theoretically possible (Bénard, 1901; Rayleigh, 1916; Knopoff, 1964) depending on the dimensions of the convective layer. The Rayleigh number ($R = g\alpha\beta h^4/\kappa\nu$ where α is the coefficient of thermal expansion; h is the depth of the layer; κ is the thermal diffusivity; ν is the viscosity; β is the temperature gradient; and g represents the gravity) for the whole mantle would be 10^{6-8}. This value is far in excess of the value of the critical Rayleigh number ($\sim 10^3$) for thermal convection in a liquid layer. This general situation is unchanged when the effects of sphericity, and rotational motion of the earth are taken into account (e.g., Chandrasekhar, 1961). When h is taken as the thickness of the upper mantle, the value of R becomes comparable with the critical value.

There are two types of models of convection in the mantle. One, the more usual, is the flow caused by vertical temperature differences, and the other is the flow driven by horizontal temperature differences (Pekeris, 1935). Bénard-Rayleigh convection belongs to the first type. Recently, however, it has been demonstrated that the hexagonal cells observed by Bénard were produced by the temperature variation of surface tension rather than by thermal convection (Block, 1956; Pearson, 1958).

Many objections have been raised against the mantle convection current hypothesis. Since the actual state of the earth's interior is supposed to be remote from that in mathematical models, many of the objections against simplified hypotheses are legiti-

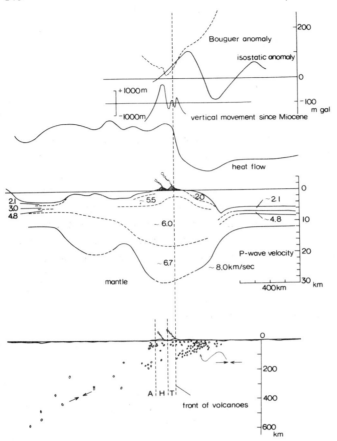

Fig.97. Typical cross-section of the east Japan arcs, compiled from Fig.11 (Bouguer anomaly), Fig.13 (isostatic anomaly), Fig.80 (vertical movement), Fig.43 (heat flow), Fig.9 (topography), Fig.17 (crustal structure), Fig.25–27 (earthquake foci), Fig.40 (types of magma: T = tholeiite, H = high alumina basalt, A = alkali-basalt) and Fig.128 (seismic stress axis).

mate. In the actual earth, physical properties, such as flow law and thermal diffusivity, are expected to have a strong layered structure and the heating must be internal as well as from below. Chemical segregation and reactions of many kinds will accompany the convective motions. But, because of these complexities, the authors consider it unwise to deny the possibility of convection current on account of simple-minded objections, especially when there are numerous supporting indications. For the island arc tectonics, we would not necessarily require such a large-scale regular mantle-wide current system as those advocated by, for example, Runcorn (1962) and Vening Meinesz (1962) and disputed by, for example, MacDonald (1963) and Lyustikh (1967).

The hypothesis of mantle convection current dates back to the 19th century. But probably Holmes (1929, 1965), Vening Meinesz et al. (1934), Griggs (1939) and Vening

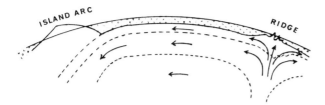

Fig.98. Schematic cross-section of the flowing upper mantle.

Meinesz (1964) among others were pioneers in putting the hypothesis forward as the basic mechanism of various phenomena, in particular of continental drift, mountain building and island arc formation. Equivalence of the average heat flow in the continent and ocean (see p.62) was taken as an indication of convection currents in the oceanic mantle (Bullard et al., 1956). Later, the high heat flow over the mid-oceanic ridges indicated that the current may be rising underneath the ridges (Rikitake and Horai, 1960; Von Herzen and Uyeda, 1963; Langseth et al., 1966). The hypothesis of ocean-floor spreading (Dietz, 1962; Hess, 1962; Vine and Matthews, 1963; Vine, 1966) is the remarkable synthesis of otherwise unrelated pieces of information into a scheme based on the mantle convection hypothesis.

Where does the current go down? Most convectionists suggest that the current goes down in the trench areas at the margin of continents (e.g., Griggs, 1939; Wilson, 1963, 1965; Orowan, 1965; 1967). We also hold this view, i.e., the current under the Pacific area rises at the East Pacific Rise and descends at the continental margin along an inclined surface dipping landward as outlined in Fig.98. It was once thought that the northwest Pacific Rise (Shatsky Rise) or the Emperor Seamounts Ridge may be the upwelling. But these topographic uplifts are inactive both seismically and thermally (Vacquier et al., 1967a) and, therefore, are rather unlikely sources of the spreading sea-floor. Now, it appears that the present rising part is the area of the East Pacific Rise, although it is so far away from Japan (see Fig.2). This view postulates that the horizontal extent of the convection cells in the mantle, the size of the Pacific Ocean, is playing an important role. Simple application of the Rayleigh theory on the onset condition of convective motion suggests that the vertical scale of the cells must be comparable to the horizontal one. However, if we assume the existence of an extremely "anisometric" flow pattern (Elsasser, 1963, 1966) over the whole Pacific Ocean, the spreading of the Pacific floor from the East Pacific Rise may be possible without invoking mantle-wide cells. We are going to examine, in the rest of Chapter 3, the possible consequences of such a system of currents and see whether or not it can adequately explain the observations in island arcs.

As noted above, the current system may not be a regular mantle-wide one, considering the non-uniformity of the mantle property (Takeuchi and Hasegawa, 1965; McConnell, 1965; McKenzie, 1967; Kanamori, 1967a, b). Recently, Takeuchi and Sakata (1970) examined the possible pattern of thermal convection cells in a Newtonian viscous laye

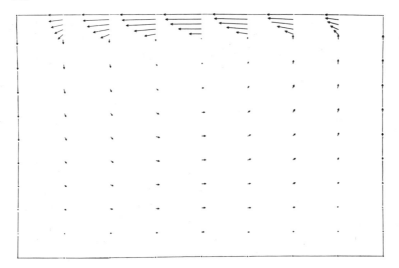

Fig.99. Velocity vector diagram in a fluid layer having variable viscosity. In this case, the upper 1/10 thick layer has the viscosity of 1/1000 of the rest of the layer. (After Takeuchi and Sakata, 1970.)

with a depth-dependent viscosity. It was assumed that the viscosity in the upper thin layer is less than that in the lower layer, implicitly simulating the mantle. They applied Rayleigh's theory of marginal stability to a two-dimensional problem and found that the stream lines are concentrated but not closed in the upper thin layer as shown in Fig.99. The aspect ratio of the critical perturbation is not so different from that in a homogeneous layer case, when referred to the thickness of the composite system. This means that the aspect ratio referred to the thickness of the upper thin layer is very large. The flow pattern.is quite similar to the "anisometric" current pattern of Elsasser. Foster (1969), on the other hand, investigated a time-dependent and finite amplitude thermal convection in an internally heated layer of fluid having a variable viscosity. When the viscosity increases exponentially with the depth and the Rayleigh number is large ($\sim 10^8$), convection was found to be confined to the low viscous upper layer and cells are shortened in the horizontal scale. Apparent disparity of the conclusions derived by Foster (1969) and by Takeuchi and Sakata (1970) would be important in considering the earth's mantle. Even with the "anisometric" current system confined in the upper mantle or a flattened cell pattern associated with the convection due to internal heating (Tritton and Zarraga, 1967), the observed smallness of the lateral scale in the topographic and heat flow cross-sections of island arcs (Fig.97) is still difficult to explain in terms of a simple Newtonian viscous fluid; the surface heat-flow distribution over a convective cell of a Newtonian fluid has a width comparable to the horizontal scale of the cell (Okai, 1959; Shimazu and Kohno, 1964).

In this respect, the hypothesis of Andradean plasticity in the mantle by Orowan (1965) or that of the descending semi-rigid lithosphere may be more appropriate. The Andradean body is characterized by a power law:

$$\dot{e} = B\,\sigma^n \tag{8}$$

where $\dot{e}$ is the strain rate, σ the stress, B a constant which includes temperature dependence, and n the exponent larger than unity. The characteristic of such a flow is the marked non-linear concentration of shear in a thin layer, providing a possible explanation of the observed sharpness in the topographical and heat flow cross-sections.

It can now be clearly seen that the establishment of flow law of the mantle is most urgently required. Rocks, as other solids, exhibit non-elastic properties under stress especially at high temperature (Jaeger, 1964). Such an inelastic deformation is called transient creep. After a sufficiently long time, the rate of deformation becomes constant. Such a process is called steady-state creep. In the mantle convection hypothesis, it is the steady-state creep with extremely low rate of deformation ($\sim 10^{-14} - 10^{-16}$ /sec) that plays the essential role. Amongst various possible mechanisms for creep in solids, Gordon (1965) and McKenzie (1968) pointed out that the thermally activated migration of atoms and vacancies from one grain boundary to another would be most important under the mantle conditions. The process is called diffusion creep or Herring–Nabarro creep. In this case the flow law is expressed by the linear Newtonian relation, i.e.:

$$\dot{e} = B'\,\sigma \tag{9}$$

where B' is the coefficient of viscosity that critically depends on temperature. On the other hand, Weertman (1970) argues that Herring–Nabarro creep should be dominant only at stresses too low for the mantle ($< 10^{-2}$ bars) and dislocation creep which gives the power law should be the dominant process. Creep experiments at low strain rates ($10^{-4} - 10^{-7}$ /sec) have only recently become possible (Griggs, 1967). According to these experimental works (Carter and AvéLallemant, 1970; Raleigh and Kirby, 1970) the power law (equation 8) with $n = 5$ rather than equation 9 appears to predominate, i.e.:

$$\dot{e} = A\exp\left(\frac{-BT_m}{T}\right)\sigma^5 \tag{10}$$

where A and B are constants and T_m is the melting temperature. Raleigh and Kirby suggest that the power law creep occurs at least in the upper 130 km of the mantle. Below this depth, there are at present too many uncertainties to decide which law holds.

Mathematical or hydrodynamic theories of thermal convection have been developed mainly on the critical perturbation for the onset of convection in Newtonian fluid. In the case when the Rayleigh number is much in excess of the critical value, finite amplitude flow becomes important. Studies on finite amplitude convection have been advanced by a number of investigators (Okai, 1959). Analytically, however, the problem is extremely

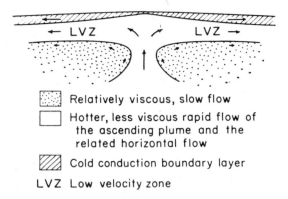

Relatively viscous, slow flow

Hotter, less viscous rapid flow of
the ascending plume and the
related horizontal flow

Cold conduction boundary layer

LVZ Low velocity zone

Fig.100. Convection model for ascending flow under mid-oceanic ridges. (After Turcotte and Oxburgh, 1969.)

difficult because of the non-linearity involved in the basic equations. Recent advancements have been made by the use of finite difference methods which became feasible owing to the advent of high speed computers. For instance, Fromm (1965) obtained solutions for a wide range of Rayleigh numbers from critical to 10^7. The flows with finite amplitude are characterized by two dimensionless numbers; one is the Rayleigh number and the other is the Prandtle number defined as $Pr = v/\kappa$.

It appears possible that flow characteristics in finite amplitude convection are quite different from those at marginal stability conditions, as hinted by the already mentioned discordant conclusions of Foster (1969) and Takeuchi and Sakata (1970). It seems that, as the Rayleigh number becomes higher, stream lines of the flow become more concentrated in the boundary region, leaving the more or less isothermal central core. On such a boundary layer model, Turcotte and Oxburgh (1967) and Oxburgh and Turcotte (1968) discussed the mantle convection and the surface heat flow. More recently, Turcotte and Oxburgh (1969) extended their arguments on finite amplitude convection to the case when the mantle has variable viscosity. In this argument, the uppermost boundary layer immediately underlying the ocean floor is taken to be cold enough to be approximated by a rigid plate, of which thickness starts from zero at a mid-oceanic ridge axis and increases as the current flows away from the axis and cools. Their model of mid-oceanic ridges shown in Fig.100 seems to be fairly realistic at the present stage.

Descending flow or lithospheric slab under trench–arc systems

As shown in the last part of Chapter 2, the subsidence of the Japan Trench seems to have taken place in the relatively recent past in geological history and perhaps it is still taking place at present. Evidently such a subsidence has been possible only when the volume of the crustal or upper mantle material has been reduced, or when the original material has been otherwise removed. Shrinkage due to local cooling, or to some phase changes to denser forms, is hardly possible for the required amount of reduction of volume. Olivine—spinel transition in the mantle is supposed to take place at a depth of about 400 km (Akimoto and Fujisawa, 1968; Fujisawa, 1968). Transition at such a great depth would not produce such a localized subsidence as that found in oceanic trenches. Basalt—eclogite transition, that may be expected at shallower depths, would also give rise to a reduction of volume by about 10%. In order to result in a subsidence of 3 km, transition of a column of 30 km basalt and, therefore, a local descent by about 30 km of the isotherm in 2—3 m.y. is required. Deserpentinization of serpentinized peridotite may be another possibility for reducing the volume (about 30%), provided that the water can be removed. In this case the $500°C$ isotherm must rise some 10 km locally in the area (Bowen and Tuttle, 1949). But it would be hard to conceive that simple conduction of heat would produce such a rapid and local displacement of isotherms. Here, one should remember that heat flow in the trench area is subnormal.

Recourse must be made, therefore, to the second possibility, i.e., the removal of material or the descending flow. If the flow goes downward in a certain area, the upper layers would be dragged down to form a trench. If indeed the light crustal material is downbuckled into heavier mantle, Vening Meinesz—Kuenen's tectogene with negative gravity would be formed. Although the seismic refraction data (see Fig.17) show no indication of downward protrusion of the crust under the Japan Trench, the crust in the deep trough is certainly thicker than in the ocean basin area, and the gravity shows a pronounced negative isostatic anomaly. In the southern Kurile region, the crust seems to have a real bulge (Fig.19). If deserpentinization or basalt—eclogite transformation of descending material takes place, accompanying reduction of volume would enhance the subsidence. The downward bulging of the crust appears to be situated under the landward slope of the trench, causing the landward displacement of the low gravity belt mentioned on pp.17—24. This landward displacement of mass deficiency toward the deeper level demands the oblique descending of the material as shown in Fig.98.

Since mechanical stresses depend on the rheological properties of the material as well as the boundary conditions, it may not be simple to estimate quantitatively the stress distribution under island arcs with a descending mantle flow. Qualitatively, however, the following conjecture may be possible. Although tensional stresses may appear locally, the predominant stresses due to the thrust along the upper surface of the descending flow would be shearing, giving rise to many shallow earthquakes with a thrust-type radiation pattern under the outer arc, i.e., the inner side of the trench.

In Fig.128 (p.196), the directions of the shear stresses are indicated schematically on the basis of the result of statistical studies by Aki (1966). The deeper earthquakes in this figure will be discussed later (see p.190), but the shallower ones are considered here. Aki's statistics of distribution of initial motion demonstrate that either the thrustings of the continental block over the oceanic block or those of the oceanic over the continental, or

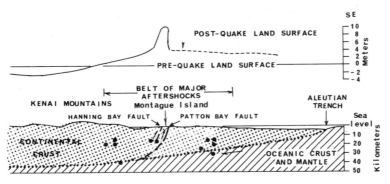

Fig.101. Schematic section across the Aleutian Arc, showing approximate (solid lines) and inferred (dashed lines) land level changes, after-shock distribution and geological interpretations. (After Plafker, 1965.) (Copyright 1965 by the A.A.A.S.)

both, are much more abundant than the faultings of other types. The authors suggest that the type of earthquakes in this area belongs to the former, that is, to the thrust of the continental block over the oceanic block, as indicated in Fig.128. Fedotov et al. (1961) arrived at the same conclusion for the Kurile Arc also from the radiation pattern of earthquakes. Further, Plafker (1965) attributes the origin of the tectonic deformation associated with the 1964 Alaskan earthquake to a major low-angle thrust faulting, in which the continental crust overlies the oceanic crust (Fig.101). This model was later substantiated by Savage and Hastie (1966) and supported by source mechanism studies by Stauder and Bollinger (1966) and Kanamori (1970). These earthquakes in Japan, Kurile, and Alaska are all situated in the outer belts of island arcs and are expected to occur under the shear stress of the equivalent directions. We call this type the outer-belt type earthquakes. The major part of seismic energy in island arcs is provided by the outer-belt type of great earthquakes of shallow origin, the mechanism being probably dynamically related to the descending oceanic lithosphere.

The Miocene alpino-type deformation was assigned to this belt (the Off—Sanriku—Zyoban geosyncline) analogous to the situation on Timor Island in the Indonesia Arc (see p.107). Could it not be allowed to expect the similar deformation under the present sea-floor? The authors infer on the basis of the shallow seismic activity mentioned above, that younger sediments in this belt are now being compressed, intensely folded and thrust. These deformations would be producing a structure similar to that in the cores of the Alpine mountain system. Our proposal on the equivalence of the low-angle thrustings in the alpino-type deformation with the outer-belt type earthquakes may be justified in view of the direction of stress as well as the highest dynamic energy concentration in the geosynclinal belt. There is no room for doubt about the importance of the role of water for the alpino-type deformation proposed by Hubbert and Rubey (1959, see also Holmes, 1965). The thrusting and folding of this type should occur relatively easily beneath the sea-floor in general. If one accepts this view, it may be quite understandable that Stille

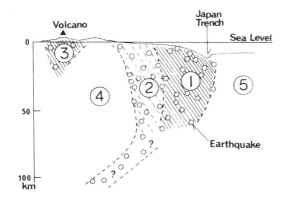

Fig.102. Schematic section across the Japan Trench showing the seismic activity and the structural state. Shaded areas represent fractured zone.
1 = seismically active and highly fractured;
2 = seismically active, but slightly fractured;
3 = moderately active and appreciably fractured (this shallow zone is beneath the volcanic belt);
4 = with no shocks and probably unfractured; and
5 = seismically inactive and homogeneous in structure. (After Mogi, 1967.)

(1955) declared that only the germano-type orogenies were known in geologically recent times. His statement should be restricted to the geology on land. If the alpino-type folding is in action on the present earth at all, it should be in the active outer-belt of island arcs beneath the sea. It is an interesting subject to try to check the hypothesis proposed here, that the pile of outer-belt type earthquakes forms the geological structure observed in the Alpine orogenic belt. It may be ascertained, for example, by seismic reflection profiling.

As stated above, in the upper layers of the Pacific island arcs, a great many shallow earthquakes are occurring. This suggests that the mechanical energy supplied by the flow in the substratum is expended partly in the form of seismic energy. Isacks et al. (1968) attribute the cause of shallow earthquakes to the bending of the oceanic lithosphere. But, the marginal part of the continental lithosphere facing the sinking slab seems to be suffering from more intense effects of the flow than the slab itself. The sea-floor should be downwarped to form the bottom of the trench and the sediments should be compressed, folded and thrust. Mogi (1967) divides the part of the crust underneath the east Japan arcs into five zones parallel to the Japan Trench on the basis of the seismic features, including regional variations of after-shock activity (Fig.102). The seismically most active and highly fractured region (*1* of Fig.102) coincides with the outer Off—Sanriku—Zyoban folded zone. The zone seems to be the most strained corner of the continental mass by the inclined descending flow of the oceanic lithosphere (*5* of Fig.102). Most of the great earthquakes in the Japanese area occur off Honsyu Island in the Pacific, i.e., the area of the Off—Sanriku—Zyoban folded zone and its equivalent in southwest Japan. It was pointed out (Takeuchi and Kanamori, 1968; Mogi, 1970) that

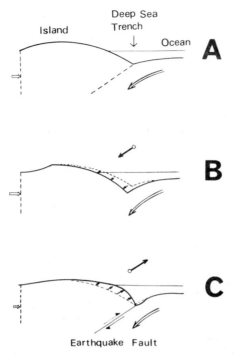

Fig.103. Assumed tectonic movements under island arcs. B. Before a large earthquake. C. At the time of a large earthquake. (After Mogi, 1970.)

the crustal deformation in Japan during inter-quake periods and those accompanying earthquakes are concordant with the model of descending flow under the arc as schematically shown in Fig.103. The descending flow exerts a drag force along the dipping zone so that the Pacific coast subsides and, in a horizontal direction, displaces landward during the inter-quake period as shown in Fig.103B. During a large earthquake, the crustal movements take place in the direction of release of the accumulated stress (Fig.103C). When the deformation due to the dragging action reaches a critical point, elastic rebound, i.e., an earthquake occurs. The crustal deformations at the time of large earthquakes have been amply observed in Japan in the last several decades. The most conspicuous one is the sudden upheaval of the order of one metre of capes on the Pacific coast (Sugimura and Naruse, 1954). Mogi (1970) has shown that these observations as well as the horizontal movements (Fig.34) can be accounted for by the model mentioned above. This model, however, has a defect that the secular uplift of the capes cannot be accounted for.

The descending current would be compatible with the observations of the low heat flow (see p.63) in the outer zone of the arc. But, the observed low heat flow seems to be too much localized in a limited zone to be caused by a Newtonian flow. As will be referred to later in this chapter (p.162), the two types of metamorphism (see p.53) may now be in progress under the high and low heat flow regions, respectively. The pressure—temperature conditions for the high p/t type metamorphism were presumably first brought about by the descending convection current. The metamorphism, being endothermic, must have absorbed the heat locally and helped to bring about the sharpness of the low heat flow zone. The virtually negative value of the Moho heat flow in the low heat flow region (Fig.44) may be accounted for by the loss of heat due to metamorphism (Uyeda and Horai, 1964).

Thanks to the recent developments in the sea-floor spreading hypothesis, the concept of mantle convection has become widely accepted. But it should be noted that the very concept had been maintained for a long time by the investigators of island arcs; e.g., A. Holmes, F. A. Vening Meinesz, D. Griggs and many other pioneers. In order to explain the classical features such as topography and gravity anomaly, it would suffice that the mantle flow goes down under trenches. Owing to the modern developments in various disciplines in the solid earth sciences, it may now be possible to be more specific about the nature of the flow dynamics of the upper mantle.

One important factor is the notion of the semi-rigid lithosphere. The uppermost 50—100 km thick layer of the earth has long been considered to be moderately strong and rigid because of the long sustained deviation from isostasy (e.g., Dietz, 1961). The upper mantle low-velocity layer on the other hand suggested that the layer immediately below the lithosphere may be relatively weak. Thermal consideration of the upper mantle (e.g., Lubimova, 1958), and rheological investigation of post-Glacial uplift (e.g., Takeuchi and Hasegawa, 1967) fully supported the general concept of a rigid lithosphere overlying a soft low-velocity layer. Deeper in the mantle it was shown that probably both rigidity and viscosity increase with the depth. The surface boundary layer model of Turcotte and Oxburgh (1968) described in the previous section harmonizes with the lithosphere concept. The possibility that the lithosphere is divided into blocks which move relative to each other without significant internal deformation was noticed by several investigators (McKenzie and Parker, 1967; Morgan, 1968; Le Pichon, 1968). The possibility of reconstruction of an ancient super-continent by the fit of coast lines, regular pattern of oceanic magnetic lineations, distributions of occurrences of earthquakes, orogenic zones, trench and arc systems and so forth all supported the above view. These authors tested this view on quantitative grounds and showed, indeed, that lithospheric plates are produced at mid-oceanic ridges and displaced rigidly relative to one another. The plates are supposed to be destroyed at the trenches. A rather important consequence of this view is that not only the continental drift but also the orogeny in general may be interpreted as due to the interaction between neighbouring lithospheric blocks. This is the notion of *the plate tectonics*. At the moment, the present authors see a good possibility that the tectonics of island arcs developed here can well be incorporated with the more general framework of the plate tectonics. In the following discussions in this chapter, however, they will try to maintain the inductive approach based on their own observations on island arcs rather than to take the deductive one from the general scheme.

An important finding in support of the plate concept in recent studies is the existence of an inclined slab of high velocity and low attenuation under active island arcs due to Utsu (1967; Fig.24) and Oliver and Isacks (1967; Fig.53). The situations shown in these figures are highly indicative of the underthrusting or descending of the lithosphere. Some results of explosion seismology (Fig.54–56) suggest also the existence of the descending oceanic lithosphere.

The lithospheric plates are driven by some forces. In the mantle convection theory, plates are considered to be passively carried by the current in the substratum. Recently, however, the hypothesis that the motion of plates is caused by their own weight (Elsasser, 1967) appears to acquire more support: plates start to sink when their density becomes large enough, due, for example, to cooling. Descent of a plate into the astheno-sphere produces a mantle earthquake zone and possible pressure-induced phase changes during the descent accelerate the motion. The rest of the plate is pulled by the leading part that sinks. This way of motion, in spite of its apparent novelty, may also be called a type of convection, which is highly non-linear. Perhaps one of the great differences between the more conventional view and this novel view would be that the velocity of the fluid motion directly below the plate is greater than that of the plate motions in the former case and smaller in the latter. For the tectonic theory of island arcs, the mechanics of the plate motion are highly important and some clue may be obtained from the mechanism of earthquakes. This point will be taken up again (p.190).

TEMPERATURE FIELD UNDER ISLAND ARCS

Temperature in the crust and upper mantle

Generally, the temperature T is governed by the equation of heat transfer:

$$\rho \cdot C_p \cdot \frac{\delta T}{\delta t} = \nabla \cdot (\kappa \cdot \nabla T) + A - \rho \cdot C_p (V \cdot \nabla T) + \sigma_{ij} \frac{\delta V_i}{\delta x_j} \tag{11}$$

where ρ is the density, C_p is the specific heat, κ is the thermal conductivity, A is the rate of heat production, V is the velocity, and σ_{ij} is the viscosity stress tensor. Because of so many unknown factors in equation 11, an estimate of temperature in the earth's interior is extremely difficult. If a steady state is assumed, the temperature in the uppermost layer, where convection is negligible, is given by:

$$\nabla \cdot (\kappa \cdot \nabla T) + A = 0 \tag{12}$$

Generally, the thermal conductivity κ consists of two or more parts with different origins, i.e.:

$$\kappa = \kappa_{ph} + \kappa_{rad} + \kappa_{exc} \tag{13}$$

Where κ_{ph} is the phonon conductivity, κ_{rad} is the radiative conductivity and κ_{exc} is the excitonic conductivity. κ_{ph} decreases with temperature as $1/T°K$ and κ_{rad} is expected to increase with temperature (Clark, 1957), but not as rapidly as $(T°K)^3$ (Fukao et al., 1968; Fukao, 1969). Little is known about κ_{exc} in the earth (Lubimova, 1963). Experiments conducted so far indicate that κ_{rad} becomes effective only at temperatures higher than, say, $1,000°C$ (Kawada, 1967; Kanamori et al., 1968). Fukao (1969) showed that up to about $1,000°C$, the thermal conductivity of olivine is best approximated by a constant, because of the cancelling effect of temperature dependence of κ_{ph} and κ_{rad}.

If κ and A are constant, the heat flow and temperature at depth z are given by:

$$Q(z) = Q(o) - Az \tag{14}$$

and:

$$T(z) = T(o) + \int_o^z \frac{Q(z)}{\kappa} dz = T(o) + \frac{Q(o)}{\kappa} z - \frac{A}{2\kappa} z^2 \tag{15}$$

When κ and A are not constant, $T(z)$ and $Q(z)$ can be computed likewise by numerical integration of equation 12. The temperatures thus computed from equations 12 and 15 may be called the "steady conduction temperatures". These equations are sufficient to give the general aspects of the temperature distribution in the uppermost layers in the earth. For instance, $T(z)$ is expected to be higher under oceans than under continents, because the A-value in equation 15 is lower for the sea than for the continent. Taking typical continental (Precambrian shield) and typical oceanic structures, the temperature distributions in the upper part of the earth were estimated by Clark and Ringwood (1964) as shown in Fig.104 by the curves PC and O. Although it must be borne in mind that considerable uncertainties are inherent in these estimations, the temperature at 100 km depth is about $300°C$ lower under the Precambrian shield than under the ocean basin. Difference in temperature of $300°C$ near the fusion temperature can be expected to cause considerable contrasts in the mechanical properties of the upper mantle under continents and oceans. Difference in the nature of the low-velocity layer under continents and oceans may be explained by this difference in temperature.

One may expect from the above consideration that in the area of transition between continent and ocean, the upper mantle temperature has a horizontal heterogeneity, the isotherms rising toward the ocean as shown in Fig.105 (MacDonald, 1963, 1965; Pollack, 1967). Such temperature distribution in continental margins appears to have penetrated into the minds of many people, as if it were generally the case. For instance, some discussions on the seismicity of island arcs as due to thermo-elastic stresses (Lubimova and Magnitzky, 1964; Pollack, 1967) are based on such a temperature distribution. But, one must not overlook that in the case of island arc type continental margins, the equivalence of heat flow in oceanic and continental areas is probably not true. Temperature distribution as shown in Fig.105 may be true only for non-active continental margins, such as the east coast areas of the American continents. This important point will be restressed below (p.162).

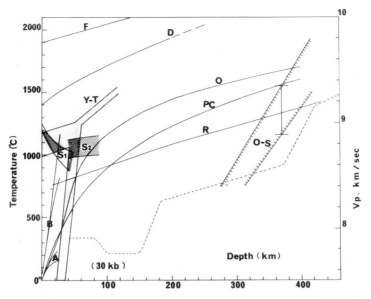

Fig.104. Various estimates of the temperature in the upper layers of the earth.

O = under oceanic areas (Clark and Ringwood, 1964);

PC = under Precambrian shield (Clark and Ringwood, 1964);

R = from electrical conductivity (Rikitake, 1952);

$O-S$ = from olivine-spinel transition; hatched zone indicates the possible range of $p-t$ of the phase transition (Akimoto and Fujisawa, 1968);

A = under the low heat flow area in northeast Japan (Uyeda and Horai, 1964);

B = under the high heat flow area in northeast Japan (Uyeda and Horai, 1964);

F = melting curve of forsterite (Davis and England, 1964);

D = melting curve of diopside (Boyd and England, 1963);

$Y-T$ = melting range of basalt (Yoder and Tilley, 1962);

S_1 = melting of basalt with $P = P_{H_2O}$; and

S_2 = melting of basalt with 1.6% H_2O (Shimada, 1966a, b).

The lower broken curve is the distribution of P-wave velocity in the upper mantle. (After Kanamori, 1967b.)

The mantle temperature has also been estimated, independent of equation 15, from the distribution of the electrical conductivity in the mantle. The latter can be inferred from the time variations of the geomagnetic field through the theory of electromagnetic induction (e.g., Rikitake, 1966). One of the estimates along this line by Rikitake (1952) is shown in Fig.104 by the curve R. The estimated temperature R is lower than PS and O by 200–300° C. Another indication of the actual temperature comes from the consideration of the olivine–spinel transition. Combining the experimental study of the transition (Akimoto and Fujisawa, 1968) with the seismometric evidence (Kanamori, 1967b), Fujisawa (1968) states that the temperature at 375 km depth under Japan should be 1,300° C ± 100° C. This value is in good agreement with R in Fig.104.

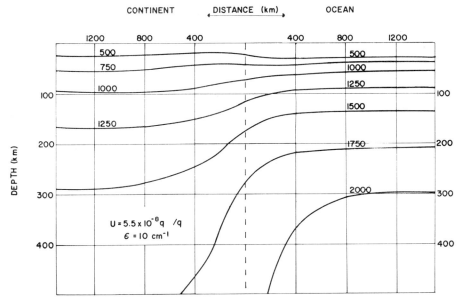

Fig.105. An example of temperature distribution under continental and oceanic structure. (After MacDonald, 1965.)

If convection current exists in the upper mantle, as we postulate, heat transport due to mass transfer becomes important in equation 11 and the approximation by equation 15 is no longer valid except at very shallow depth. In an actively convecting fluid, the temperature gradient can be approximated by the adiabatic gradient:

$$\left(\frac{\delta T}{dZ}\right)_s = \frac{g \cdot \alpha \cdot T}{C_p} \tag{16}$$

where α is the coefficient of thermal expansion. Putting reasonable values in equation 16, the gradient for the upper mantle is estimated to be 0.2–0.4° C/km. This value is considerably lower than the usual values of the temperature gradient inferred from solid conduction equations. Thus, if convection exists, temperature tends to be lowered. The difference between O and PS and R might be partly related to the above situation. The rigid lithosphere may be the region where equation 15 holds and the low viscous substratum is the region where the adiabatic gradient is realized, corresponding to the boundary layer model of Turcotte and Oxburgh. Thus, the T–Z curve tends to have a rather sharp bend at about the lower boundary of the lithosphere. Perhaps this is the depth where the actual temperature approaches closest to the melting temperature.

Indeed, once one accepts the view that the mantle is in convective motion surface heat flow over the earth must be interpreted anew and used as a check of theories. Many investigators (e.g., Langseth et al., 1966; McKenzie, 1967b; Turcotte and Oxburgh, 1969;

Sclater and Francheteau, 1970) have shown that the observed heat flows over mid-oceanic ridges and ocean basins are essentially compatible with the general postulate of the new global tectonics: heat flow profiles have been computed on various models and compared with the observations. In this way, it is now considered that the observed equality of the mean oceanic and continental heat flows may be entirely fortuitous and no use in deriving important conclusions with regard to the thermal history of the earth. At the present stage, it must be fully realized that all the thermal history considerations based on computations of solid conduction equations are very remote indeed from the actual earth's history.

We now turn to the question of melting temperatures. The usual procedure of estimating the melting point for the whole mantle is to derive the melting curve from seismic data (Uffen, 1952; Gilvarry, 1957). This approach would be inadequate for our purpose, because the estimated curve does not seem to have sufficient precision, in particular at shallow depths. Extrapolation of experimental melting curves of particular materials (Yoder and Tilley, 1962; Boyd, 1964) may be more reliable in some cases. McConell et al. (1967) made a study of the melting behaviour of "pyrolite", which is believed by Green and Ringwood (1963) to be the main constituent of the mantle. Some of the melting curves are shown in Fig.104. If there is some water in the upper mantle, melting behaviour in an aqueous system must be important. It has been shown that melting temperature of basalt (Yoder and Tilley, 1962; Shimada, 1966a, b) and peridotite nodules (Kushiro et al., 1968) falls by several hundred degrees when H_2O is present (see, for example, curves S_1 and S_2 in Fig.104). The melting temperature which is of the greatest importance in magma generation would be the solidus of the mantle material. The melting curve of basalts may be close to it. As noted in the preceding paragraph, the estimated actual temperatures in the upper mantle seem to approach the melting temperature in the depth between 70–200 km. In general, elastic wave velocities are increasing functions of pressure and decreasing functions of temperature. Thus, 6–14°C/km has been calculated for the minimum temperature gradient to produce a low-velocity layer (e.g., Anderson, 1966). Fig.104 also shows an example of velocity–depth relation for P-waves in the upper mantle (Kanamori, 1967b). The velocity is lowest in the depth between 100 and 200 km. It would imply that the temperature gradient at this depth interval is high, and the temperature is close to the fusion temperature. Kanamori (1967a) suggests that the low Q-value of the low-velocity layer can be explained by assuming that some 5% partial melting is actually taking place there.

Temperature under island arcs and its implications

Much of the argument on the upper mantle temperature (p.156) applies to normal areas. In island arc areas, where heat flow values vary drastically in zones, the subterranean temperature distribution would also be widely varied (see p.63). In fact, the

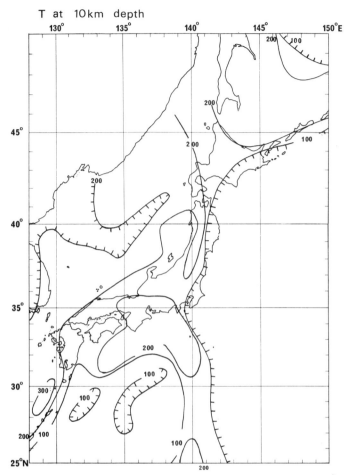

Fig.106. "Steady conduction temperature" at 10 km depth. (After T. Watanabe, personal communication, 1970.)

temperature beneath the low heat flow zone would be as low as 150–200°C at the 20 km depth whereas it would be almost 700–800°C beneath the high heat flow zone as shown by the curves *A* and *B* in Fig.104. These curves are the "steady conduction temperatures" under the Japanese Islands (Uyeda and Horai, 1964). In applying equation 12 to the first approximation, *A* was regarded as consisting entirely of radiogenic heat. Then taking the crustal thickness as shown in Fig.22 and assuming the mean representative content of radioactive elements of crustal rocks, the temperature and the heat flow at various depths were computed from the observed surface heat flow values. As to κ, a reasonable model was adopted based on the experiments by Birch and Clark (1940). The same type of calculation was extended to the entire Japanese area by taking the average heat flow values for each 1°–squares (T. Watanabe, personal communication, 1970). In his calculation, the crust was divided into granitic and basaltic layers with a thickness ratio of 2:1 for the continental area while the crust is assumed to be entirely basaltic in the oceanic area. The distribution of the "steady conduction temperature" at 10 km depth is shown in Fig.106. The pronounced contrast in the temperature between the outer and inner zones of the island arc should be noted.

The temperature distribution in the mantle under the Japanese area may be characterized by a high temperature zone inside the volcanic front and a low temperature zone outside it. At the depth of 10 km (Fig.106), for example, the difference in the temperature under the Japan Sea and the western Pacific is expected to be more than $100°$ C and the difference is expected to increase with depth. At the depth of 20 km the difference in the "steady conduction temperature" seems to exceed $600°$ C. The situation is clearly in contrast to that under a normal continental margin (see Fig.105); therefore, the geophysical discussions on continental margins based on the latter type of temperature distribution are not to be applied to the island arc type margins. Discussions on thermo-elastic stress or upper mantle electrical conductivity anomalies of island arcs must take the above temperature distribution into consideration.

Curves A and B in Fig.104 indicate that the temperature beneath the low heat flow zone is much lower than the solidus temperature of earth materials, while it approaches to the solidus under the high heat flow zone. Thus, we can expect that no magmas are allowed to exist under the low heat flow zone whereas magmas are likely to exist under the high heat flow zone. The general distribution of active volcanoes in the Japanese Islands (Fig.38) must be related to this situation.

Because of the fact that the observed regional metamorphism in Japan took place before the start of the orogenic activity related to the present island arcs, the meta-morphic belts described in Chapter 1 (p.53) are not the members of the present island arc features. However, the same types of metamorphism as those of the older ages may be in progress at present beneath the island arc regions (Miyashiro, 1959, 1961b). As shown in Fig.37, three pairs of metamorphic belts are exposed in the Japanese Islands. The youngest pair is composed of Hidaka and Kamuikotan metamorphic belts, lying parallel to each other. The metamorphism that is related to the next oldest orogeny in Japan produced the pair which is composed of the Ryoke—Abukuma, or simply Ryoke meta-morphic belt, and the Sanbagawa metamorphic belt. The regional metamorphism of the Sanbagawa belt is glaucophanitic, producing glaucophane, lawsonite and jadeite. This metamorphism is supposed to have taken place under a relatively high pressure and low temperature, or under a high p/t condition. On the other hand, the regional metamor-phism in the Ryoke terrain produced andalusite, cordierite and sillimanite in rocks of appropriate compositions and are accompanied by a group of batholiths of granitic composition. The Ryoke belt is characterized by a low-pressure and high-temperature type metamorphism, i.e., a low p/t type. The Hida and Sangun belts are the oldest of the three pairs of metamorphic belts. Here again the outer Sangun belt and inner Hida belt are represented by the high p/t and the low p/t metamorphism. As noted before (pp.53

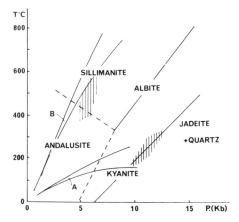

Fig.107. Stability relations of andalusite, kyanite and sillimanite, and of the jadeite–quartz assemblage and albite. The shaded areas represent the $p-t$ fields for the high p/t and low p/t types of regional metamorphism. (After Miyashiro, 1961.) Curves A and B are the "steady conduction temperatures" for the high heat flow and low heat flow of Japan. (After Uyeda and Horai, 1964.)

and 94), it is common in the circum-Pacific regions that the high p/t-type and low p/t-type metamorphic belts lie parallel to each other. Takeuchi and Uyeda (1965) examined the possibility that the two types of metamorphism may be in progress under Japan as a result of the island arc activities in a quantitative manner. They postulated that it is very likely that the two types of regional metamorphism are under way beneath the zones of low and high heat flows in the east Japan arcs. Fig.107 is the summary of their conclusion: the temperature–depth curves computed by Uyeda and Horai (1964) for the low and high heat flow zones in Japan (curves A and B in Fig.104) go through the temperature–pressure fields of the two types of metamorphism inferred from the petrological studies (Miyashiro, 1961a). However, it appears unlikely that the geosynclinal sediments were actually buried to the depth exceeding, say, 35 km, corresponding to the pressure greater than 10 kbar in Fig.107. For high p/t type metamorphism, the deviatoric stress caused by the drag of the descending slab must have played an important role in raising the effective pressure at shallower depths (Kumazawa, 1963).

The temperature distribution should influence the mechanical properties. In fact, the delay in the arrival time of seismic waves in the volcanic zone (Fig.23) may be at least partly due to the anomalously high temperature. More remarkable are the recent findings on the heterogeneity of the upper mantle under island arcs as referred to on pp.32 and 156. The high Q and high V zone dipping toward the continent and the relatively low Q and low V zone under the inner arc (Utsu, 1966; Oliver and Isacks, 1967; Kanamori and Abe, 1968; Fig.24) are in harmony with the temperature field proposed here.

Another important feature that may probably be related to the temperature field is the upper mantle electrical conductivity anomaly (see p.70). When the earth's magnetic field undergoes external disturbances, such as magnetic bays, daily variations or magnetic

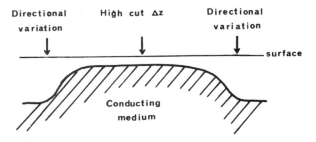

Fig.108. A. Symbolic representations of subterranean distribution of high conductive layer for "High cut ΔZ" type and "Directional variation" type anomalies. (After Uyeda and Rikitake, 1970.) B. Normal continental margin (a); and island arc type continental margin (b). (After Uyeda and Rikitake, 1970a.)

storms, electric currents are induced in the earth and these currents in turn produce secondary magnetic fields. An observer at the earth's surface should feel the sum of the primary (external) and secondary (internal) fields, and the latter should depend on the distribution of the electrical conductivity in the earth. The electrical conductivity of the mantle is expected to depend on temperature, pressure and composition. If other features are the same, the temperature effect should dominate. Here, tentatively, we take this view, so that high conductivity is taken as corresponding to high temperature. According to experiments, the conductivity of olivine, for instance, increases exponentially with temperature. The high conductive layer in the following discussion may be taken, to the first approximation, as representing the 1,200–1,500°C isotherm. According to the theory of the electromagnetic induction (e.g., Rikitake, 1966), when a highly conducting (therefore hot) layer is placed close to the surface, high frequency components of ΔZ (vertical component of the variation field) should be surpressed. This simple type of anomaly is observed, actually, for instance, in Iceland (Hermance and Garland, 1968). On the other hand, when the depth of the upper limit of the conducting (hot) layer varies geographically, it is expected that the vector of the variation field is controlled internally, in the manner that time-dependent magnetic lines of force are rejected by the conductor. When the isotherms are raised more under the ocean than under the continent, Parkinson's vector (Parkinson, 1959, 1962) tends to point to the ocean. Typical examples are found in Australia (Parkinson, 1964) and on the west coast of the North American continent (Schmucker, 1964; Coode and Tozer, 1965). Uyeda and Rikitake (1970) called the former type "high-cut ΔZ type" anomaly and the latter "directional variation type" anomaly as illustrated in Fig.108A. Nagata (1967), aiming at a similar effect, called them "frequency anomaly" and "vector anomaly", respectively.

There seem to exist two cases of the "directional variation type" anomalies associated with the continent–ocean boundaries: i.e., the normal continental margin type and the island arc type as shown symbolically in Fig.108B. In the symbolic profiles of sea coasts in this figure, the seawater is also indicated, because of its high electrical conductivity. As mentioned on p.157, temperature–depth relations in continental margins, calculated on

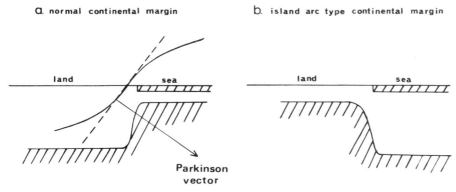

a. normal continental margin   b. island arc type continental margin

Parkinson vector

Fig.108 B.

the assumption of the equivalence of heat flow in the continent and ocean, indicate that the subterranean isotherms, and therefore the equi-conductivity surfaces, are supposed to be as shown by a in Fig.108B. The "directional variation type" anomaly associated with such a case will be such that Parkinson's vector tends to be directed toward the ocean. This may be called the normal ocean effect. When the heat flow in the ocean is higher than in the continent, the anomaly of ΔZ of this type would be enhanced. However, in the case of active island arcs like the Northeast Honsyu Arc, heat flow is subnormal in the ocean and high in the continent side, making the distribution of subterranean temperature entirely reverse to the ordinary type as shown by b in Fig.108B. In such a case one would not observe the normal ocean effect. Usually, geomagnetic variations have a larger north—south than east—west component at low and moderate latitudes; therefore, any subterranean structure elongated in the north—south direction should result in only small anomalous changes. Northeast Japan is an essentially north—south trending feature. For the above two reasons, the well known Central Japan Anomaly (Fig.47) is not the type expected for the north—south trending coast of the Northeast Honsyu Arc. It has been proposed recently, however, that the Central Japan Anomaly may be partly due to the normal ocean effect of the east—west trending Pacific coast of the southwest Honsyu Arc rather than of the north—south trending one of the Northeast Honsyu Arc (Rikitake, 1969; Uyeda and Rikitake, 1970). The rest of the anomaly may be accounted for by the effect of the ocean water around the peninsulas, on which most of the magnetic observatories are located (Sasai, 1968). One important fact is that the heat flow in the Sikoku Basin is not subnormal (see p.65), but there are numerous high heat flow values. It can, therefore, be expected that the southwest Honsyu coast exhibits a normal type of the ocean effect. The Central Japan Anomaly which has long been a mystery may be explained in this way. As mentioned above, not much "directional variation type" anomaly can be expected for northeast Japan. The Northeast Japan Anomaly (Fig.47) is considered to be caused by a rather complex temperature distribution. The depth of mantle conducting layer (i.e., the 1,200—1,500°C isotherm surface on our model)

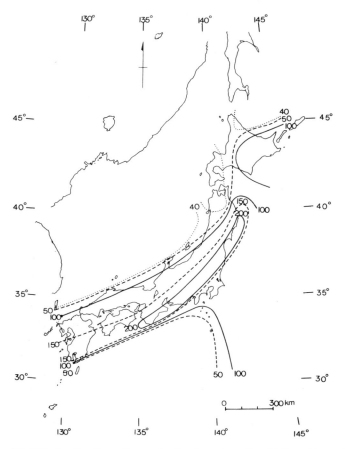

Fig.109. Depth of mantle conducting layer in units of kilometres under the Japanese Islands. (After Rikitake, 1969.)

deduced mainly from the conductivity anomaly is shown in Fig.109. The compatibility of this temperature distribution with that deduced from heat flow is remarkable.

A situation similar to that of northeast Japan can be expected for the Pacific coast of South America where heat flow is expected to be low in the trench area and high in the Andes where there are volcanoes (Uyeda and Watanabe, 1970). In fact, it has been reported that, in contrast to the North American coast, no normal ocean effect has been detected near the Peruvian coast (Schmucker et al., 1964), and that the effect of upraised high conducting mantle ("high-cut ΔZ type" anomaly) exists in the Andes (Forbush et al., 1967).

One of the most remarkable features of island arcs is that the mantle earthquake zones are closely related in space to the volcanic belts, i.e., the latter roughly coincides with the 120–200 km isobaths of the mantle earthquake zone (Fig.27 and 38). The existence of

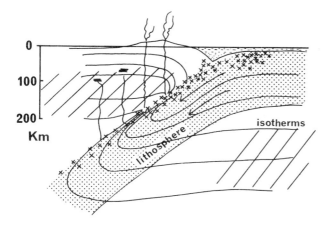

Fig.110. Probable thermal structure under active island arcs.

the well defined volcanic front seems to suggest strongly that there are some critical conditions for production of magma under the island arcs. In an attempt to explain these close relationships, the authors propose the hypothesis that, to the first approximation, magmas under island arcs are generated at the depths where the upper mantle is sufficiently heated by the deep and intermediate earthquakes. As will be restated later in this chapter, this hypothesis is based upon the idea that deep earthquakes are an "exothermic process" and play the role of local heat sources under the island arcs: the upper surface of the descending lithosphere begins to be heated up by the deep earthquakes and at the critical depth of 120–140 km actual melting starts. This depth defines the volcanic front and the layer below this depth may be the low-velocity layer. In other words, the magmas are produced along the seismic plane where it intersects the upper mantle low-velocity layer. The general situation described here may be summarized as in Fig.110. From the heat flow distributions, we infer that isotherms in the mantle under the northwestern Pacific Basin are much deeper than under the continental side of the island arc. As the cool oceanic lithosphere sinks down on approaching the island arc, the isotherms are more depressed. Endothermic reactions, such as dehydration of serpentine and sedimentary rocks, may accompany the sinking and enhance the depression of isotherms. High activities of shallow earthquakes are expected to take place near the corner, where the lithosphere is forced to descend, and its continental side, where the alpino-type deformation is taking place. After the complete change of the direction of motion, earthquakes are expected to occur only along the boundary between the sinking mass and the subcontinental mass, as indicated by the "seismic plane" of deep earthquakes. Many investigators now seem to consider that deep earthquakes occur in the middle of the lithospheric slab, rather than at its upper surface (e.g., Utsu and Okada, 1968; Isacks and Molnar, 1969; Fig.24). This view comes mainly from the result of inference based on the mechanism of the deep shocks (see p.198), but not from seismometric evidence, because the seismometric observations are as yet unable to determine the relative positions of the lithosphere and mantle earthquake zone accurately enough. The friction causes the local heating and eventually the production of magma. The magma thus produced will ascend more or less vertically and raise the isotherms under the inner side of the arc. Under the Sea of Japan area, depth of magma production is so great that magma cannot reach the surface, but can contribute to the regional high heat flow. Along the lower surface of the lithosphere, there is no seismicity, possibly because the lower layer of the asthenosphere is ductile. It is also suspected that the lithosphere is carried passively by the flow of the asthenosphere, so that there is no relative movement between the lithosphere and the asthenosphere.

Thermal process of sinking lithosphere and a possible origin of the Sea of Japan

As has been explained above, the thermal regime under island arcs is probably highly complicated. Numerous attempts (Langseth et al., 1966; McKenzie and Sclater, 1968; Turcotte and Oxburgh, 1968; McKenzie, 1969; Minear and Toksoz, 1969) have been made to account for the low and high values of heat flow in island arc areas on the model of the descending lithosphere. Some possible mechanisms of production and transfer of heat have been proposed by these authors, such as viscous dissipation, latent heat related with metamorphism or phase transformation, and heat transfer due to convection current or penetrative upward movement of magmas. Here we will describe the work by Hasebe et al. (1970) which examined the model outlined in Fig.110. This paper will be abbreviated as the H.F.U.-paper below.

There is no direct observation about the velocity of the descending lithosphere along the deep seismic zone. In the H.F.U.-paper, the velocity was assumed to be 3 cm/year. The velocity is an important factor in the thermal regime and its effect is being examined by the authors of the H.F.U.-paper since various values have been suggested for various arcs (Brune, 1968; Le Pichon, 1968 or Table X). An important part, however, of the arguments below, which are based on the assumption that the slip rate is 3 cm/year, will be valid irrespective of actual numerical value of the slip rate as long as it remains at the same order of magnitude.

The calculations were made on a two-dimensional model of which the cross-section is simulated to that of the northeast Honsyu Arc, which has the curved deep seismic zone dipping with an inclination of about 1: 3 toward the continent. The geophysical observations on this section are summarized in Fig.97. It was assumed that a 100 km thick lithosphere descends along the seismic zone (Fig.111). Between the descending lithosphere and the upper mantle above it there would be a shear stress σ. The upper mantle except for the lithosphere is treated as essentially fixed.

For the lithosphere to move with a velocity v against the shear stress σ, the work W

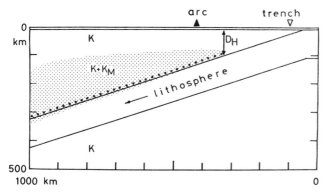

Fig.111. Schematic representation of the model in the H.F.U.-paper. Uppermost double line shows the 5-km crust and asterisks show the region of "frictional" heating. Dotted zone represents the occurrence of partial melting. Other symbols are explained in the text.

must be done per unit area along the shearing boundary per unit time, without depending on the mechanism of the relative movement. Thus:

$$W = \sigma v \qquad (17)$$

Little of W will run away as seismic waves (Aki, 1967), but most of W will take the form of thermal energy, i.e., the "frictional" heat. This situation will prevail below a depth, D_H in Fig.111, which is greater than some tens of kilometres where the rocks would lose their ordinary brittleness. The relation between v and σ depends on the rheological properties, i.e., viscosity etc., of the mantle and descending lithosphere. In the H.F.U.-paper, the value of W was given as a parameter, without going into the detail of the physical mechanism of its generation. At the lower boundary of the lithosphere the gradient of the velocity field was considered to be far smaller than at the upper boundary. Because of small shear stress at the lower boundary, the work done there would be small and ignored. The depth at which "frictional" heating starts, D_H, is also treated as a parameter in actual calculations. The crust was considered to be basaltic and to have a thickness of 5 km. The normal temperature distribution for the oceanic basin was taken as the initial condition.

The materials of the lithosphere and upper mantle beneath the island arc—trench system may contain some amount of H_2O, which would play an important role in the melting behaviour of the mantle material (Kushiro, 1966). It was assumed in the H.F.U.-paper that the material involved with the process of magma production under island arcs is saturated with water, so that a model solidus curve was synthesized from the works of Yoder and Tilley (1962) and Kushiro et al. (1968a).

As mentioned above, the descending lithosphere and the upper mantle on it are heated by a kind of "frictional" heat. When the temperature exceeds the solidus temperature of the constituent material, magma will be generated. The magma rises rapidly and when it arrives at a place where the temperature is lowered to the solidus its upward movement stops. With the ascending magma some heat will be transported advectively but after its solidification heat will be conducted only by the normal solid conduction process. To be strict, the physical mechanism of magma ascent must be considered. The authors of the H.F.U.-paper, however, only try to estimate the necessary amount of heat transportation without going into details of the process. Namely, the effect of massive transfer of heat in the partially molten mantle is represented by a parameter with a dimension of thermal conductivity.

The equation of thermal conduction accompanied by movement of material and internal heating is as follows (the equation is the same as equation 11, except that A includes every possible heat generation):

$$C_P \cdot \frac{\delta T}{\delta t} = \nabla \cdot (\kappa \cdot \nabla T) + A - C_P \, (\vec{V} \cdot \nabla T) \tag{18}$$

This equation was transformed into a finite difference equation of two dimensions and solved numerically through the alternating direction implicit method.

In order to estimate the amount of the heat transportation by the ascent of magma in the partially molten portion, an increased effective *vertical* thermal conductivity κ_E was introduced as:

$$\kappa_E = \kappa + \kappa_M \tag{19}$$

(κ_M corresponds to the heat transportation by the movement of magma).

The boundary conditions were as follows: at the top and the bottom the temperature is fixed, at the side from which the lithosphere comes into the system the temperature is fixed and at the opposite side the horizontal heat flow is fixed to zero. The length of a time step Δt and the grid aperture Δx is taken to be 0.5 m.y. and 10 km, respectively, in most cases. Other constants were taken as usual, i.e., specific heat per unit volume and normal thermal conductivity were 1 cal/cm^3 degree and 0.01 cal/cm sec degree, respectively. In actual models, "frictional" heat W was expressed by the amount of heat produced per unit vertical column H_E. Remembering the way W is defined, H_E and W are related by:

$$H_E = W/\cos \alpha \tag{20}$$

where α is the inclination angle of the descending lithosphere. Thus A in equation 18 is expressed in terms of H_E, which is conveniently given in the heat flow unit (H.F.U.). Varying the parameters, κ_M, H_E and D_H, the heat flow patterns at the surface of the earth at present, which was tentatively assumed to be about 100 m.y. after the beginning of the descent of the lithosphere, were computed. Comparing the computed results with the observed heat flow pattern, the proper values of the magnitudes of these parameters may be estimated.

Some results of computations are shown in Fig.112—114. The parameters for different

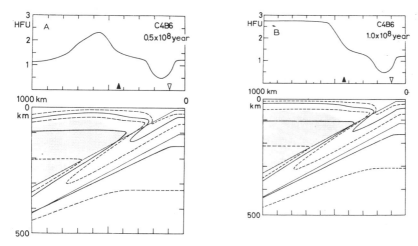

Fig.112. Heat flow pattern and temperature distribution of model C4B6. A. After 0.5·10⁸ year. B. After 1.0·10⁸ year. Shaded area represents the partially molten region. The symbols ▽ and ▲ express trench and the front of volcanic zone, respectively. The solid lines are 500°C, 1,000°C and 1,500°C isotherms. The broken lines show the 250°C, 750°C and 1,250°C isotherms.

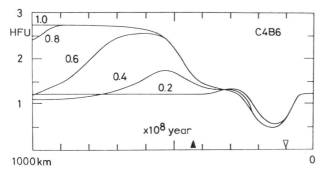

Fig.113. Variation of the heat flow pattern of model C4B6 at every 0.2·10⁸ year.

models are listed in Table IX. The model for which the pattern of the heat flow after 100 m.y. matches the observation best was C4B6 (Fig.112, 114). The parameters of this model are H_E = 6.0 H.F.U., κ_M = 0.10–0.08 cal/cm sec degree and D_H = 60–80 km. At the points corresponding to the Sea of Japan the computed heat flow is about 2.5 H.F.U. These characters coincide with the observation very well. The "partially molten" zone in Fig.112 reaches to about 50 km depth. One of the remarkable features is that the point at which heat flow is highest is located above the deeper heat source and not right above where the "frictional" heating is assumed to start. Fig.113 shows the variation of heat flow pattern of model C4B6 at every 0.2·10⁸ year. The high heat flow region covers almost all of the Sea of Japan after 100 m.y. and a stationary state is attained over most of the region concerned. Another notable aspect is that in spite of the very high heat flow

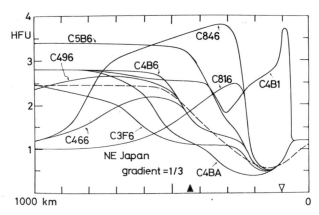

Fig.114. Heat flow patterns of model C4B6 and other models, after 10^8 year (broken line shows the observed heat flow pattern).

TABLE IX

List of model descriptions

Model	H_E (10^{-6} cal/cm^2 sec)	κ_M (0.01cal/cm sec degree)	D_H (km)	Δt (m.y.)
C3F6	4.5	14	60	0.5
C466	6.0	5	60	0.5
C4B1	6.0	10	10	0.5
C496	6.0	8	80	0.5
C4B6	6.0	10	60	0.5
C4B6	6.0	10	100	0.5
C816	12.0	0	60	0.5
C846	12.0	3	60	0.5

of 2.5 H.F.U., the temperature of the mantle is kept rather low. For instance, the highest temperature appearing in Fig.112 is 1,700°C at 300 km depth near the left boundary. The temperature in the heated zone is 300–500°C higher than normal. These situations are due to the assumed increase of κ in the molten region. In the descending lithosphere itself, the temperature gradient is inverted. The negative gradient in the slab reaches about $-10°$C/km. This gradient is about three times as large as that of the upper side of the descending lithosphere. This negative gradient means that a large amount of heat is flowing downward there, manifesting the heat sink nature of the descending lithosphere.

Now, let us examine some other models. When $H_E \leqslant 4.5$ H.F.U., the heat flow is generally too small (See C3F6 in Fig.114 and Table IX). It may be suspected that heat flow will be made higher if a larger κ_M is employed. But this is not the case, because the heat sink effect of the sinking slab does not allow the local temperature to reach the solidus when $H_E \leqslant 4.5$ H.F.U., so that equation 19 cannot be used. When $H_E \geqslant 7.5$ H.F.U., on the other hand, the heat flow becomes too high (C845 and C5B6 in Fig.114).

If κ_M is made lower, the peak value of heat flow may be lowered, but the thermal time constant for the deeper heat source becomes enormous, so that the surface heat flow in most of the Sea of Japan after 100 m.y. can hardly be affected. This effect can be seen by comparing the models C846, and C816 in Fig.114. The same tendency can be seen for H_E = 6.0 H.F.U. cases (C4B6, C496, C466 in Table IX and Fig.114). When κ_M = 0.05 cal/cm sec degree, the high heat flow region is too narrow. Thus, the parameters are constrained to a narrow range. In order to have a spread of the high heat flow region in 100 m.y. as observed, κ_M must be greater than 0.08 cal/cm sec degree. Thus, the parameters are constrained to a narrow range. It can easily be seen that D_H affects the surface heat flow distribution in the area between the trench and the island (compare C4B6 with C4B1 and C4BA). The most suitable value of D_H would be about 60 km.

As described above, the values of two parameters H_E and κ_M, which are needed to explain the observation of surface heat flow in the Sea of Japan region, were estimated to be 6.0 H.F.U. and 0.08–0.10 cal/cm sec degree, respectively. These results may have the following significance.

First, let us examine the significance of the amount of heat H_E.

(1) We first evaluate the importance of the latent heat of phase transformations and the temperature rise due to adiabatic compression. From the temperature distribution obtained in the model C4B6 (Fig.112), it may be inferred that serpentine would be stable to the depth of several tens of kilometres if a sufficient amount of water is present in the descending lithosphere. The upper mantle, however, may be undersaturated, i.e., $P_{H_2O} <$ P_{load} (Kushiro, 1966; Ringwood, 1966; Kushiro et al. 1968a) so that it is difficult to evaluate the depth of stability exactly. Moreover, by the lack of thermodynamical data for the dehydration reactions, great uncertainty cannot be avoided in the estimation of latent heat of the reaction taking place at depths. Uyeda and Horai (1964) calculated the required heat for the dehydration of serpentine to be about 400 cal/cm^3, which is based on the heat absorption of about 20 to 30 kcal/mole of water. The amount of serpentine can be as much as 1/20 of the descending lithosphere, i.e., an equivalent thickness of 5 km in the 100 km thick descending lithosphere. This would correspond to the case where the third layer of the oceanic crust is considered to be serpentinite, as in the serpentinite crust model of Hess (1955). A column of descending material with a thickness of 100 km and unit cross-section would require a heat of about 400·10·(1/20) cal/cm^2 during the time of its passage through the system which is about 30 m.y. in H.F.U. models, the equivalent average heat flow being −0.2 H.F.U. If Ringwood's pyrolite model is assumed, the ampholite (olivine + amphibole) may be considered stable in the descending material from the stability field proposed by Ringwood (1966) and the temperature gradient of the normal mantle in our model. Kushiro (1966) and Kushiro et al. (1967) suggested that the possible hydrous minerals in the upper mantle were amphibole and mica (phlogopite) and the maximum depths of their stability were 100 km and 150–200 km, respectively. Heat of dehydration of these phases, however, would be much smaller than the above figure for serpentine because of their lower abundance. Then the total heat required in

the dehydration in the descending slab would be less than, say, one-fifth of the normal surface heat flow.

Minear and Toksöz (1969) calculated the heat generation in the basalt–eclogite and olivine–spinel phase transitions. Their values were about 1.5 cal/g and 80 cal/g, respectively. For the olivine–spinel transition some new experimental data were given by Ito et al. (1969), in which the entropy change, Δs, was −4.4 cal/mol. degree. Using this value and temperature of about $1,000°\,C$, the heat generation accompanying the transition is estimated to be about 37 cal/g for a Fo_{80}–Fa_{20} olivine. This corresponds to an equivalent heat flow of about 1.6 H.F.U. if the same method of estimation as in the case of dehydration of serpentine is applied.

The temperature rise due to adiabatic compression is given by equation 16. Consider a column of descending lithosphere as in the case of the estimation of heat for dehydration, then, the temperature rise of this column by adiabatic compression from the top to the bottom of the mantle region under consideration is calculated by:

$$\int_{top}^{bottom} \frac{dT}{T} = \int_{top}^{bottom} \frac{\alpha \cdot g}{C_p}\, dz \tag{21}$$

Taking $\alpha = 2$–$3 \cdot 10^{-5}$/deg, $g = 10^3$ cm/sec^2, $C_p = 0.3$ cal/g and the descent during 30 m.y. from the top to the bottom to be about 300 km, we get T (bottom) = T (top) $\times$ 1.06. Thus the temperature rise is found to be about $40°\,C$, assuming that the mean temperature of the column when it was at the top of the upper mantle was $500°\,C$. We can reduce this result to an equivalent heat generation in this 100 km high column by the relation:

$$H \sim C \cdot \Delta T \cdot D/\Delta t = 0.5 \text{ H.F.U.} \tag{22}$$

where D, C, ΔT, and Δt are the thickness of the slab, the specific heat in unit volume, temperature rise, height of the column and time required for its 300 km descent, respectively.

In order to estimate the true value of "frictional" heating H_{EO}, such heat sources as discussed here must be subtracted from H_E: since the heat due to phase changes and that due to adiabatic compression are estimated to be 1.4 H.F.U. and 0.5 H.F.U., respectively, the resultant H_{EO} becomes about 4.1 H.F.U.

(2) We now consider H_{EO}, which is supposedly due to the "frictional" heating W per unit area along the shearing surface. H_{EO} and W are related by the definition $W = H_{EO}\cos \alpha = 3.9$ H.F.U. McKenzie and Sclater (1968) estimated the stress heating along the shearing plane as $W \sim 0.5$ H.F.U. Although this value is comparable with the observed value of the heat flow anomaly itself, they showed that this heat cannot reach the surface

within the time of about 300 m.y. as long as the upper mantle behaves as a solid with a thermal conductivity of 0.01 cal/°C cm sec. They also indicated that their estimate of 0.5 H.F.U. is uncertain by an order of magnitude. Turcotte and Oxburgh (1968) estimated the heat generation by viscous dissipation from the diffusion creep theory, and obtained a value of $W \sim 1.4$ H.F.U. These values are all rather small compared with the estimate in the H.F.U.-paper, and seem to be insufficient to explain the surface excess heat flow because of the heat-sink effect of the descent of the slab. McKenzie (1969) discussed, however, that the difficulty could be overcome by the stress heating that would take place at a depth much closer to the earth's surface than along the deep seismic plane. Minear and Toksöz (1969) also estimated the effect of the strain heating as well as that of other heating such as the phase transformation and adiabatic compression. Their results showed that strain heating may be comparable in magnitude with other heating processes. The obvious consequence is that it would be difficult to explain the observed heat flow anomaly by their models.

It is, indeed, important to note that the descending cold plate is a great heat sink. In other words, a considerable portion of the heat produced in and near the descending materials would be spent for self-heating. The heat production of 1—2 H.F.U., which is equivalent to the estimates by Turcotte and Oxburgh (1968) and McKenzie and Sclater (1968) can hardly be observed at the surface. In fact, in H.F.U. models, the mean temperature of the lithosphere is about 500°C at the top when it enters the system and about 1,100°C (model C4B6) at the left edge when it is leaving the system. The heat lost downward from the system per unit time can be expressed as $Q = \Delta T \cdot C \cdot v \cdot D$, where D is the thickness, v the velocity of the slab and $\Delta T = 1,100° - 500° = 600°C$. Considering the length of the slope, $L \sim 900$ km, this loss of heat can be represented by an equivalent heat flow of about -4.4 H.F.U. ($=Q/L$). This value is consistent with the heat budget in H.F.U. computation: in model C4B6, H_E, the radioactive heat and the surface heat flow were 6.0 H.F.U., 0.9 H.F.U. and 2.0—2.7 H.F.U., respectively. Most of the difference must be carried down into the mantle by the sinking lithosphere.

(3) What will be the magnitude of the shear stress, σ, from which the probable value of the work done W may be calculated? From the general relation $W = \sigma \cdot V$, the magnitude of σ is estimated as 2 kb to produce W of 5 H.F.U. It appears that this value of σ is an order of magnitude too great for the usually accepted value of stress drop of earthquakes. On the other hand, the value of viscosity, η, may be estimated as follows; assuming the thickness, a, of the shearing layer to be a few tens of kilometres, $\eta = W \cdot a/v^2 \sim 1.8 \cdot 10^{22}$ poise. In the H.F.U.-paper, heat generation was estimated through numerical experiments. Whether or not such an amount of heat production is physically possible remains to be clarified in future studies.

Secondly, the significance of the enhanced thermal conductivity, κ_M, will be as follows:

(a) The relation between the effective thermal conductivity κ_M and the heat transport by the ascending magma may be obtained approximately. The heat flux Q_M owing to the ascent of magma through the partially molten layer is expressed in the actual calculation as:

$$Q_M = \kappa_M \cdot \Delta T / \Delta Z \tag{23}$$

where ΔT is the temperature difference between the top and the bottom of the layer, and ΔZ is the thickness of the layer.

On the other hand, the heat transport by the magma flux V, i.e., the volume of magma ascending through unit horizontal area per unit time, is:

$$Q_M = C \cdot V \cdot \Delta T \tag{24}$$

Therefore:

$$V = \kappa_M / C \cdot \Delta Z \tag{25}$$

As can be seen in Fig.111, ΔZ is about 50—200 km. Therefore, V may be obtained as $V = 0.6—0.14$ cm/year. On the other hand, the value of V calculated from the amount of erupted volcanic material during the Quaternary Period in Japan is: $V = 0.002$ cm/year according to Sugimura et al. (1963). These figures are compatible when a few percent of the magma participating in the heat transport in the upper mantle actually appear at the earth's surface.

(b) If the magma continues ascending at the rate of, say, 0.3 cm/year, the total thickness will reach 300 km in 100 m.y. A room size of the upper mantle under the Sea of Japan must be provided to accommodate the ascending material. This result is not inconsistent with the idea that Sea of Japan was formed by an opening (Tokuda, 1926; Terada. 1934; Murauchi, 1966; Karig, 1969, 1970; Nakamura, 1969). The present Japanese Islands originally were the eastern marginal parts of the Asiatic continent, but after the commencement of the process envisaged above, the Japanese Islands had to drift southeastward as shown in Fig.115. Furthermore, the total amount of the ascending magma during the time that is taken by the descending lithosphere to reach a depth of 300—500 km is computed to be about 100—150 km. This would mean that an amount of the same order of the descending lithosphere itself should have risen before reaching the 300—500 km depth. In this case, most of the descending lithosphere would ascend to form the upper mantle of the area of the continental side of the arc, so that it may be considered that the rise in a way corresponds to the secondary rising mantle current behind the arc (Holmes, 1965, p.1022; McKenzie, 1969). This process may be an integral part of the seaward growth of continents at their margins (Dietz, 1961). It should be noted that the formation of the Sea of Japan by the opening due to the ascent of magmas everywhere as envisaged here may be different from the opening of, for example, the Atlantic Ocean by one large principal ridge. Such a difference would be best manifested in the pattern of geomagnetic anomalies. In a sea like the Atlantic Ocean, regular magnetic lineations would be developed through the Vine—Matthews process on both sides of the central ridge. But in the case of the Sea of Japan, there would be numerous small and short-lived

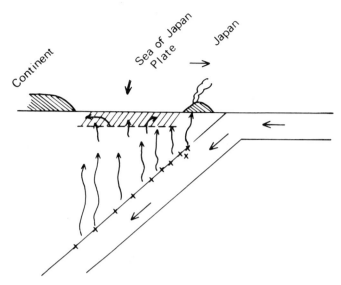

Fig.115. Possible origin of the Sea of Japan by opening.

spreading centres so that the lineations would be much less regular. Yasui et al. (1967) had actually found such a magnetic anomaly pattern in the Sea of Japan (Fig.46). Further magnetic surveys are still being done (Isezaki and Uyeda, 1972).

GENERATION AND ASCENT OF PRIMARY MAGMA

Mechanism of magma generation: non-brittle earthquakes as a possible cause of melting

As we have seen in the preceding sections, many of the island arc features seem to be accounted for by the hypothesis of descending mantle flow. However, the mechanism of generation of magma and high heat flow in the inner zone of the arc presented a difficult problem, because, unlike the ascending part of the mantle flow, the temperature of the descending part is considered to be lower than the average. The authors, thus, examine here the hypothesis that the origin of heat is the frictional heat generated through the process that causes deep earthquakes (see p.167). In the present section, this hypothesis will be explained further.

In spite of much work in seismology, what actually happens at the focus of an earthquake has not yet been made clear. Earthquakes occurring at shallow depths (say, above 10–20 km depth) are due to brittle fractures of rocks. In this case, energy is accumulated in the form of elastic strain energy until the stress surpasses the strength. Only little thermal energy is expected to be evolved in the source region before rupture and, at the moment of rupture, a sizable portion of the energy will be radiated away by elastic waves. Eventually, of course, the waves will be attenuated and their energy partly transformed into heat during propagation. After all, the heat generation associated with shallow earthquakes is not concentrated in the source region, except for some frictional heat at the earthquake faults. The latter may cause melting of the wall of the faults but it will not be extensive. On the other hand, according to the experiments (Griggs and Handin, 1960), rocks under confining pressure greater than that at 20–30 km depth in the earth are expected to be ductile and no brittle fractures occur. The shearing stress at these depths must be of the order of a hundred times the strength of surface rocks if it is to overcome the friction. The high temperature prevailing in the mantle also makes rocks non-brittle. These facts explain why most earthquakes are of shallow origin. But, in island arcs and some other areas, there are actually earthquakes of deeper origin.

Rocks may remain brittle under high pressure by the effect of pore pressure of interstitial fluid. Raleigh (1967a) postulates that some mantle earthquakes may be of brittle type, provided that such a process as dehydration of hydrous minerals takes place extensively. Possible presence of H_2O and its role in the upper mantle would be one of the most important problems, in particular of island arcs. Under island arcs, this type of earthquake may be possible to a depth of about 200 km. Below the minimum at this depth, there is often an increase of seismicity around a depth of 400 km. It may be possible that olivine–spinel and other transitions expected at these depths cause the earthquakes. Since these pressure transitions are exothermic ($\simeq$100 cal/cm^3 according to S. Akimoto, personal communication, 1968), it may well be the energy source of both seismicity and heat flow. This line of thought may be an alternative interpretation of deep seismicity under island arcs which is worth closer examination. It may, however, be difficult to produce magmas by these mechanisms.

In order to make earthquakes happen, the following conditions must be fulfilled: firstly the stresses must accumulate and secondly earth material must be able to radiate the stored energy instantaneously. The mantle under ordinary places should lack one or both of these requirements, while in the mantle earthquake zones under island arc areas *both* of the requirements are met. We propose that the upper mantle is in a convective motion, and the stress localization adequate to cause earthquakes is possible when the current meets the stationary continental mantle under island arc regions. Orowan (1960, 1965) discusses this problem and suggests that the non-brittle fracture causing deep earthquakes is due to an instability in hot creep range of the crystalline mantle. In a body with Andradean plasticity, the shear is concentrated in thin layers and displacement along such layers may give rise to earthquakes accompanying a stress drop of the order of 100 b.

This type of non-brittle fracture may produce much more heat than the ordinary brittle fracture. This heat generating nature of non-brittle fracture is only a supposition at present and the authors emphasize the necessity of its experimental verification. Here we tentatively call earthquakes of such a kind "hot earthquakes" as contrasted with the "cold (brittle) earthquakes" that occur in the uppermost layers of the earth. Griggs and

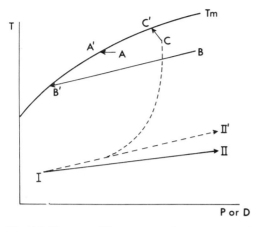

Fig.116. Three possible processes of magma generation.

Baker (1967) tried to show, through numerical experiments, that shear-melting instability due to thermally activated flow may cause both deep earthquakes and local intense heating.

There may be three possible processes for magma production in the earth (Fig.116). The first one is the change of pressure at earthquake occurrence. If the pressure decreases, the melting temperature will be lowered so that, if the temperature is already close to the melting temperature, partial melting may result (line $A-A'$ in Fig.116). But this alone can hardly be an effective process of magma production. The amount of the release of pressure in earthquake source regions cannot be greater than the "stress drop". The second process is Verhoogen's (1954) idea of producing magmas in a rising convection current which corresponds to the contact of the melting curve with the adiabatic curve (line $B-B'$ in Fig.116). Obviously the rising mantle current would be favourable for magma production, and the volcanic activity or the spreading itself at mid-oceanic ridges and rift valleys may be explained in this manner.

The magma production under an island arc, the third process, may be considered as follows: the shear strain in the downward flow of the lithosphere is concentrated in its upper surface, that is, the mantle earthquake zone (Fig.110). This concentrated strain would be the source of not only the seismic but also the magma generating thermal energy. The shearing along the focal plane would produce seismic faulting, and when the temperature of the media is high enough, it will also cause shear melting (line $C-C'$ in Fig.116). Release of stress as well as temperature rise will happen in this case, the latter probably being dominant. For this process to take place, the point C in Fig.116 should be fairly close to the solidus curve. But under the island arcs, the descent of the lithosphere is supposed to start from a point such as I in Fig.116 and the change of $p-t$ conditions during the descent should be along the adiabatic curve $I-II$, which has a much lower gradient than the melting curve. Even if we take the effect of heating by conduction from

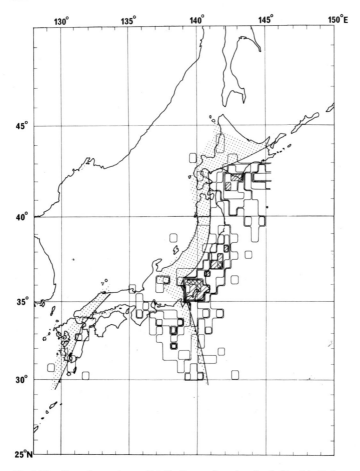

Fig.117. Complementary distribution of volcanic belts (dotted areas) and mantle earthquake epicenters. Contours show population density of mantle earthquakes deeper than 60 km during 31 years from 1926 to 1956, and are 2, 4, 6, 8, 10 and 20 shocks in the area enclosed by half-degree latitude and longitude lines. Lighter and heavier shadings represent more than 10 shocks and 20 shocks, respectively. Heavy lines represent volcanic fronts. (After Sugimura, 1966; data from the Japan Meteorological Agency, 1958.)

the surroundings in account, the curve would be like $I–II'$, and there is still little possibility of bringing the temperature nearer to the melting curve. This would be the principal problem in the present theory of island arcs. As stated already, we propose that the earthquakes in a non-brittle region are exothermic and heat the upper surface of the descending lithosphere as shown by the curve $I–C$, while the bulk of the lithosphere follows the curve $I–II'$. If this is the case, intense seismicity in the mantle earthquake zone will be favourable for magma production. This may be the reason why the upper

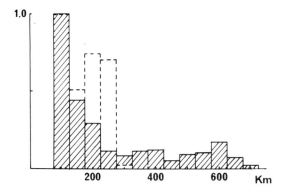

Fig.118. Ratios of frequency of shocks deeper than 125 km to frequency of 75–125 km shocks. Shaded columns show those in island arc areas at various depths and columns enclosed by dashed lines show those in non-island arc areas. Note the deficiency of shocks from 175 km to 275 km depth in island arc areas. (After Sugimura, 1966; data from Gutenberg and Richter, 1954.)

limit of the magma producing zone appears to be shallower in the seismically more active arcs (see Table X).

Energy partition of seismic waves and heat in the fracture processes is not well known so that it is hard to make progress in energetics in this respect. We emphasize again that the study of such thermo-mechanic properties of rocks under high $p-t$ conditions would be of the utmost importance. But it may be safe to assume that once melting takes place *extensively,* the very part of the mantle will no more be able to rupture, and sliding or fluid-like flow at a low rate under small shear stress will proceed and no earthquake will result. In fact, a complementary relationship in the distribution between hypocenters of mantle earthquakes and the magma-generation zone is observed as follows (Sugimura, 1966):

(*1*) The hypocenters of mantle earthquakes are abundant in the trench-side area of the volcanic front, whereas they are much less abundant underneath the volcanic belt (Fig.117).

(*2*) The numbers of mantle earthquakes at depths between 175 and 275 km relative to those between 75 and 125 km are distinctly lower in island arcs than in non-island arc regions (Fig.118). Under the island arcs magma would be generated at depths from 175 to 275 km.

(*3*) The complementary relationship can be seen also in the continent-side area of the volcanic front as illustrated in Fig.60 for the Andean Arc.

Such a complementary relation between deep seismicity and volcanism may appear to conflict with the view that magmas are produced by deep earthquakes. In our view, however, the observed negative correlation would be a natural consequence: the regional stress is more or less uniformly exerted along the mantle earthquake zone by the descending mantle current. In the areas where the melting is less extensive earthquakes are more abundant and vice versa. If the temperature is too low, the earthquakes cannot result in melting of any considerable extent. If the temperature is too close or equal to the melting point, further shear melting is also hindered because of continuous flow without earthquakes. In a rather critical range of temperature, both plastic instability and shear melting would be possible. Since the melting is expected to take place only in a confined layer, the bulk of the descending flow is generally at a lower temperature than its environs (see Fig.112). The result of Utsu (Fig.24) shows relatively small absorption along the mantle earthquake zone, and the colder upper mantle concept suspected on the basis of electrical conductivity (Rikitake, 1956) is also consistent with this idea. The part of the upper mantle regions right above the seismic or magma-producing zone, however, would be at a high temperature because of the rise of the magmas, as was explained earlier (see p.167 and Fig.112).

Phase relation and zone partial melting

It was suggested in the preceding section that magmas under island arcs are produced by mantle earthquakes. This hypothesis is supported by the fact that the volcanic belts in island arcs are associated in space with the mantle earthquake zones. Our hypothesis is not in conflict with that of Kushiro and Kuno (1963) that the basaltic primary magmas are generated in the mantle earthquake zone. Whatever the direct cause may be, the partial melting would first produce the melt with the composition at the invariant point in the system including the possible constituents of the upper mantle. Experimental studies of partial melting have been conducted by Kushiro (1965, 1968) in the pressure range up to 40 kb for the systems $Mg_2SiO_4-SiO_2-X$, where X represents $NaAlSiO_4$ and $CaAl_2O_4$. Kushiro found that in all these systems the composition of the isobaric invariant point and that of the forsterite—enstatite liquidus boundary shift from the silica-rich to the silica-poor side with the increase of pressure. Examples are shown in Fig.119 and 120. These experiments show, at least qualitatively, that the composition of the initial melt of possible upper mantle material (garnet peridotite, Ringwood's pyrolite or a similar constituent) will become less siliceous with the increase of the ambient pressure. This is in agreement with the idea of Kuno et al. (1957) that the basic differences in chemical composition in the basaltic primary magmas from tholeiite to alkaline basalt types reflect the differences in the depth of magma production, rather than the successive stages in the differentiation of a magma. Quantitatively speaking, however, the pressure, under which tholeiitic and alkaline basaltic magmas are produced, are ca. 10 kb and 20—30 kb, respectively, according to Kushiro's experiments. These pressures correspond to depths of 30—40 km and 70—100 km, which are about one third of the depth ranges assigned for the production of the magmas in the last section. This result would put the proposition that the magmas are produced in the mantle earthquake zone into difficulty.

In order to avoid this difficulty, Kushiro (1968) proposes a mechanism of partial zone melting. First, the magmas are assumed to rise more or less vertically. A probable process for such a vertical ascent would be zone melting (Harris, 1957; Shimazu, 1961;

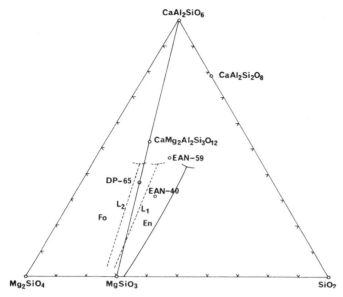

Fig.119. The forsterite (Fo)−enstatite (En) liquidus boundary at different pressures in the system forsterite (Mg_2SiO_4)−$CaAl_2SiO_6$−silica. Solid curve is a forsterite−clinoenstatite (protoenstatite) boundary at 1 atm. Dashed curves L_1 and L_2 indicate the forsterite−enstatite boundary curves at pressures near 20 and 30 kb, respectively. The compositions of DP-65, EAN-40 and EAN-59 are studied. (After Kushiro, 1965.)

Magnitzky, 1964), since at great depths no open conduit would be available for the magma. If there is a molten zone in the mantle, the temperature distribution within this zone will tend to be adiabatic because of thermal convection within the molten zone. Since the gradient of the melting point is generally much greater than the adiabatic temperature gradient, the temperature of the melt in contact with the top wall of the zone will be higher than the melting point whereas that in contact with the bottom wall will be lower. Solidification will proceed at the bottom wall releasing the latent heat and fusion at the top wall absorbing the latent heat. Through this process, the liquid state will displace upwards. Of course, since the temperature of the solid material of the roof is generally lower than the melting point, the heat energy lost in fusion of the roof is greater than the energy gained by solidification of the floor. Therefore, depending on the depth and the volume of the starting liquid, some of the magma would solidify during its upward journey. However, the basaltic magma, as stated already, appears to be the product of partial melting of the mantle material rather than its complete melt. Kushiro (1968) proposes that a process similar to the zone melting would take place for the partial melt also, and calls it zone partial melting. The partial melt will go up through the matrix of unmolten crystals, spending less energy in heating than in the case of complete zone melting. During the zone melting, volatile and alkaline components would tend to stay in the liquid, so that the volatile concentration in the magma would increase through

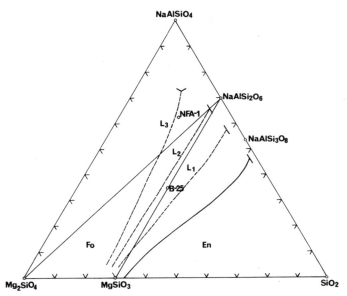

Fig.120. The forsterite (Fo)—enstatite (En) liquidus boundary at different pressures in the system forsterite (Mg$_2$SiO$_4$)—nepheline (NaAlSiO$_4$)—silica. Solid curve is a forsterite—protoenstatite boundary at 1 atm. Dashed curves L_1, L_2 and L_3 indicate the forsterite—enstatite boundary curves at pressures near 10, 20 and 30 kb, respectively. B-25, and NFA-1 are the compositions studied. (After Kushiro, 1965.)

the upward displacement. Finally, the pressure of the volatile components in the magma would exceed the ambient pressure, so that some fissures would be formed in the surroundings. Once this should happen the magma would ascend quickly through the fissures. We might call this depth the "critical depth of magma rise". Below the critical depth, the magma rises extremely slowly, maintaining its composition in equilibrium for the depth, whereas above the critical depth the rise will be much quicker and retain the composition fixed at the "critical depth". The critical depth would be determined by the balance between confining pressure and the amount of volatile concentration in the magma. The latter would be greater for a magma that has travelled a longer distance upwards. It would, therefore, be expected that the greater the depth of magma production the greater the critical depth. If such is the case, it may be inferred that the critical depths for the tholeiite magmas and alkaline basalt magmas can be 30—40 km and 70—100 km, respectively, whereas the depths of their initial melting are 130—150 km for the former magma and 250—300 km for the latter (Fig.121). We consider that this hypothetical mechanism envisaged by Kushiro may be right, though a thorough quantitative treatment is required as to the accumulation of volatile pressure during zone melting. It may be added here that the "critical depth" is by no means the same as the depth of the so-called magma reservoir in the crust.

Recently, Kushiro (1966) indicated that H$_2$O may exist in the upper mantle and may

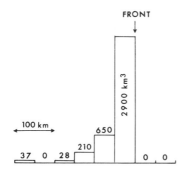

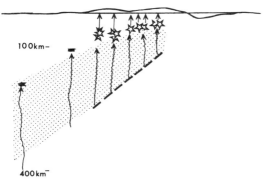

Fig.121. Cross-section across northeast Japan. Upper columns are the amount of Quaternary volcanic materials. (Revised from Sugimura et al., 1963.) Stars represent the "critical depth".

be playing some important role in the melting behaviour of the mantle. H_2O presence would not only lower the melting point, but also make the pressure derivative of the melting point (dT_m/dP) negative (Yoder and Tilley, 1962; Shimada, 1966a, b).

In the present model of island arc tectonics, the uppermost layers of the ocean bottom (the lithosphere with its sedimentary cover) are supposed to go down along the seismic plane. Therefore the material involved with magma production is likely to contain much water, so that the melting temperature may be particularly low.

Another important effect of H_2O on the melting behaviour would be the possible extension of the pressure range in which the incongruent melting of a forsterite—enstatite—silica system takes place. Kushiro et al. (1968b) have shown that the range, which is about 5 kb for forsterite—enstatite—silica under dry conditions, is likely to be extended to over 20 kb under wet conditions. This will certainly help to lessen the discrepancy between the depth of magma generation and the critical depth assumed above. Further studies are under way on this important possibility. Occurrence of andesite in island arcs is also an important problem. There have been a number of hypotheses to explain why andesite magma exist in the island arcs. To name a few, they are: the fractional crystallization of basaltic magma; the remelting of crustal materials; the assimilation with crustal materials of the basaltic parental magma; and so forth (e.g., McBirney, 1969). Recently, Yoder (1969), Kushiro (1969) and I. Kushiro (personal communication, 1970) have shown that under wet conditions with H_2O of 2% in weight or 2.5% in volume, partial melting of peridotite would produce andesite magma rather than basaltic. Kushiro maintains that under island arcs the sinking lithosphere is rich in water so that andesite is generated.

Variations across and between island arcs in volume of volcanic products, chemistry of primary magmas, and heat flow

Sugimura et al. (1963) estimated the volume of volcanic products in each 50 km wide belt parallel to the volcanic front in east Japan as shown in Fig.121, based on the same data as those used for Fig.39. The important features indicated by Fig.121 are formulated in two points:

(*1*) There is no volcano on the ocean side of the island arc, whereas abundant volcanoes exist continentward of the volcanic front.

(*2*) The regional sum of the amount of volcanic products show an almost "exponential" decrease from the front toward the continent.

The first feature is easily explained with the conspicuous changes in level of isotherms observed in Fig.110 or, more quantitatively, in Fig.112. The explanation of the second feature may be that, in our scheme, magmas are supposed to be first generated on the seismic plane dipping toward the continent and the magmas originated at greater depths will naturally have more chance of solidifying during the upward displacement. It is also conceivable that the production of primary magma itself becomes less extensive at greater depths due to the increasing departure of the actual temperature from the fusion temperature (Fig.104).

The primary magma becomes less siliceous and more alkaline toward the continent, showing a close correlation with the pattern of the isobaths of mantle earthquake foci (Fig.122). This zonality of the composition may be explained by the difference in the critical depths of magma defined in the preceding section (see p.185). Consequently, the composition of the primary magmas is most siliceous and least alkaline at the volcanic front where the critical depth is the shallowest of the same volcanic belt.

It seems probable that the amount of heat energy H_E derived from the possible descending flow differs from one island arc to another. If the heat energy supplied to the mantle earthquake zone in island arc "A" is greater than that in island arc "B", the partially molten range where magmas are generated (the hatched part in Fig.110) would be greater in "A" than in "B". The upper limit of magma-generation zone would be shallower in island arc "A", than "B". Therefore, the critical depth of magma ascent would also be shallower in "A", say 70 km deep at the front, than in "B", say 100 km deep at the front. Thus, the primary magma generated underneath the front of island arc

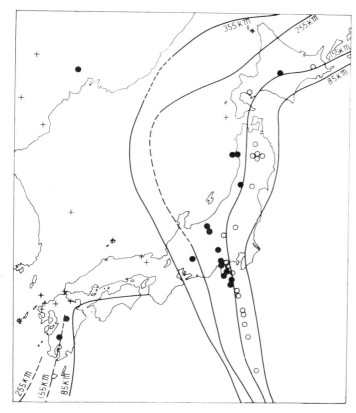

Fig.122. Petrographic provinces and isobaths of deep and intermediate earthquake foci. + = Cenozoic volcanoes derived from alkali-basalt magma; • = Quaternary volcanoes from high alumina basalt magma; and ○ = Quaternary volcanoes from tholeiite magma. (After Kuno 1960; isobath after Sugimura, 1960.)

"A" would be more siliceous and less alkaline than the magma at the front in island arc "B".

It is found that, for each island arc, the silica index (or θ-index defined on p.58) of rocks is highest at the volcanic front. The value is generally constant along the front (Sugimura, 1967b, table I). The average silica index for the volcanic front of an island arc would be taken as one of the characteristic numbers for that island arc. On the other hand, intense seismicity in the mantle earthquake zone will be favourable for magma production. These may result in the relationship that the critical depth at the front appears to be shallower in the seismically more active arc "A" than the less active arc "B". Furthermore, these differences seem to be caused by the difference in the rate of sinking movement of the oceanic slab.

Le Pichon (1968) obtained a geometrical model of the surface of the earth in terms of rigid blocks in relative motion with respect to each other. With this model a simplified but consistent picture of the global pattern of surface motion was given on the data on the sea-floor fracture zones and the zones of magnetic anomaly. In particular, the vectors of differential movement in the compressive belts were computed (Le Pichon, 1968, fig.6 and table V). His results can be related to the deep and intermediate seismicity, the critical depth of the magma producing zone at the front, and other characteristic numbers for island arcs.

An increase in the velocity of the descending current may produce the following effects:

(1) The oceanic trench will be deepened.

(2) The frequency of mantle earthquakes will increase.

(3) The maximum depth at which shear failures can occur will increase. This may result in a deepening of the lower limit of the mantle earthquake zone.

(4) The minimum critical depth of melting will decrease owing to the intensified shearing. This may result in the production of primary magma with more siliceous and less alkaline composition at the volcanic front.

These effects probably determine the positive correlation between five quantities shown in Table X. The full detail of these four items and their possible interrelationships are described elsewhere (Sugimura, 1967b).

We now turn to the problem of heat flow. Total heat discharge from the earth's interior, Q_T, consists of two parts: one is the heat carried by material transfer, Q_M, and the other is the conducted heat, Q_C, i.e.:

$$Q_T = Q_M + Q_C \tag{26}$$

In ordinary places, $Q_M \ll Q_C$ holds to a good approximation. Q_M is high in active volcanoes and geothermal areas. Q_M may be classified into the heat transported by volcanic ejecta, Q_{MV}, and by hot springs and other geothermal activities, Q_{MH}. In the unit of microcal/cm^2 sec (H.F.U., see p.62), Q_{MV} averaged for the Japanese volcanic zones is 0.04–0.3 H.F.U. (Sugimura et al., 1963) and Q_{MH} averaged over Japanese volcanic zones would be something like 0.8 H.F.U. (Fukutomi, 1961). Q_C consists of three parts, i.e., the normal heat flow, Q_{CO}, regional anomaly, Q_{CR}, and local anomaly, Q_{CL}, so that Q_C may be expressed as (Horai and Uyeda, 1969):

$$Q_C = Q_{CO} + Q_{CR} + Q_{CL} \tag{27}$$

Generally, Q_{CO} is about 1.5 H.F.U. and Q_{CR} = 0.5–1.5 H.F.U. (high heat flow zone) and–0.6 H.F.U. (low heat flow zone). Q_{CL} varies greatly in geothermal areas.

It has been noticed for some time that the heat carried out by volcanic activity is negligibly small compared with the conducted heat flow (e.g., Verhoogen, 1946). But this is true only for the average earth. In volcanic areas, the heat of volcanism may play an important role. Estimate of volcanic energy from the amount of volcanic ejecta gives, 0.04–0.3 H.F.U. for the Quaternary Japanese volcanic zone, 2.5 H.F.U. for the Quaternary Kamtchatka volcanic zone (Polyak, 1966), and 1.8 H.F.U. for Postglacial Icelandic vol-

TABLE X

Positive correlation of characteristic numbers for island arcs

Region[1]	Rate of flow (cm/year)	Deepest of the trench[2] (km)	Seismicity of the mantle[3]	Deepest of the mantle earthquakes[4] (km)	Silica index at the front[5]
Tonga (8)	9.1	11	0.19	680	42.0 ± 1.1
East Japan (3,4,5)	8.9	11	0.09	590	41.8 ± 0.3
Kuriles (1,2)	8.2	10	0.17	650	41.6 ± 0.6
Aleutian (20,21)	5.8	8	0.05	220	37.5 ± 0.5
Central America (18,19)	5.6	7	0.08	280	38.1 ± 0.6
Indonesia (36,37)	5.4	7	0.06	720	36.0 ± 0.7
West Japan		8	0.05	270	38.9 ± 0.6
U.S.A. (non-island arc area)			0.00	35	35.3
Hawaii (non-island arc area)			0.00	60	35.3

[1] The numbers are after Le Pichon (1968).
[2] Mostly after Fisher and Hess (1963).
[3] The numbers indicate the annual frequencies of the deep and intermediate earthquakes in the entire island arc with the magnitude ⩾ 7.0, per 1,000 km of the volcanic front, for 1919–1952. After Gutenberg and Richter (1954).
[4] After Gutenberg and Richter (1954).
[5] See p.58 for silica index or θ−index.

canism (Bodvarsson, 1955). It must be noted that these estimates are based on the amount of effusives only. If some volcanic materials stay in the earth as intrusives, more energy would be discharged as anomalously high heat flow, i.e., as Q_{CR} and Q_{CL} in equation 27.

Heat flow distribution in the Japanese Islands and their environs (Fig.43) shows a remarkable zonality, which cannot be accounted for by the mechanisms such as the localization of radioactivity in the crust, erosion, sedimentation or uplift of the surface (Uyeda and Horai, 1964). The main origin should exist in the mantle. As to the origin of the low heat flow Q_{CR}, an explanation by descending convection and metamorphism was given on p.167. With regard to the high Q_{CR}, one notices that the volcanoes, hot springs, and geothermal areas are located exclusively in the inner zone of the island arcs (Fig.38 and 42) and this zone is nearly exactly coincident with the zone of high Q_{CR} (Fig.43). The close correlation between high Q_{CR} and volcanism is recognized generally in other parts of the world (Horai and Uyeda, 1969).

Simple calculation shows that the regional high heat flow Q_{CR} in the Sea of Japan area cannot be accounted for by solid conduction of heat in the mantle alone. The "steady conduction temperature" in the upper mantle computed from the surface heat flow tends to become too high to keep the mantle in the solid state. Some non-conductive heat transfer mechanism may be in operation in the upper mantle under the Sea of Japan. The rising mantle convection current hypothesis was proposed to account for the high heat flow in the Sea of Japan (Shimazu and Kohno, 1964; Murauchi, 1966): Murauchi suggests that the Sea of Japan is a developed rift and that formerly the Japanese Islands were in contact with the Asiatic continent. Some Mesozoic strata in Korea are similar to those in western Honsyu and northern Kyusyu, suggesting a southward drift of Japan (Kobayashi, 1941). This hypothesis appears to be worthy of further examination in the light of other evidence. A similar but slightly different explanation for the transfer of excess heat under the Sea of Japan area would be the upward movement of magmas (see p.168), probably through the mechanism of zone partial melting as explained on p.183. Although no volcanic eruption has been observed in the Sea of Japan, there are indications of some recent volcanic activity as illustrated by Fig.123.

MECHANICAL PROCESS INFERRED FROM SEISMIC WAVE RADIATION

Stress field and possible anisotropic nature of the upper mantle

On the basis of seismic travel-time studies and physical properties of rocks, it is inferred that the upper mantle is mainly composed of peridotite, of which the major component mineral is olivine. Petrological and mineralogical considerations led Green and Ringwood (1963) to assume that the mantle is made of pyrolite, of which the major component is also olivine. Olivine belongs to the orthorhombic system and its physical properties are different in different crystallographic directions. Verma (1960) reported

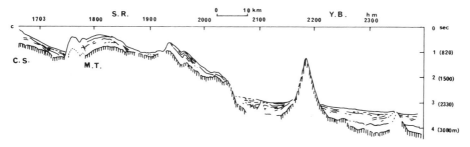

Fig.123. Seismic reflection record in the Sea of Japan, from 39°N 139°E to 39°50′N 138°E, indicating the post-depositional age of a seamount (Hotta, 1967). C.S. = Chokai Sho; M.T. = Mogami Tai; S. D. = Sado Ridge; Y.B. = Yamato Basin.

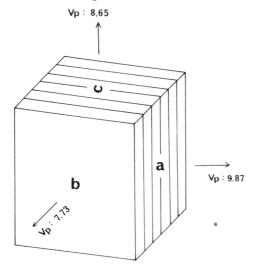

Fig.124. Orientation of principal optical directions, seismic velocities (Vp) and glide planes for magnesian olivine. (After Verma, 1960; and Hess, 1964.)

that the compressional wave velocities in the three axes are: the a-axis: 9.87 km/sec: perpendicular to the plane (100); the b-axis: 7.73 km/sec: perpendicular to the plane (010); and the c-axis: 8.65 km/sec: perpendicular to the plane (001). Verma's results may be schematically represented in Fig.124. The plane of Miller index (010) is the plane of the best cleavage.

On the other hand, olivine crystals are known to have a strong tendency to develop a preferred orientation of their crystallographic axis under directional stress (Turner, 1942; Raleigh, 1967b). If there is a flow in the mantle, therefore, the preferred orientation of olivine crystals will grow and as a result the peridotite will show an anisotropic nature in the velocity of seismic waves. H. Kanamori and H. Mizutani (personal communication, 1968) measured the P-wave velocities in three perpendicular directions of some peridotite nodules, one being taken perpendicular to the banding plane in texture, and found some 20% anisotropy.

Hess (1964) and Kumazawa and Tada (1964) independently examined the refraction seismic data taken near the Mendocino fracture zone in the area of the East Pacific Rise and over the Mid-Atlantic Ridge, respectively, and obtained the same result; that the seismic wave velocities are anisotropic in the uppermost mantle of these areas. In both cases, the maximum velocity was found to be perpendicular to the axes of the rise and the ridge and the minimum velocity was found to be parallel to these axes. That is, the direction of the minimum velocity was perpendicular to the strike-slip (transform) faults: i.e., the Mendocino fracture zone in the east Pacific and those perpendicular to the ridge crest in the Atlantic. These observations are in agreement with the idea that the (010) plane of olivine, the most easily gliding plane, is statistically oriented perpendicular to the strike of the crests and subsequently the faults develop parallel to these planes (010).

Kumazawa (1967) further studied the thermodynamics of the recrystallization process of olivine under directional stress fields and showed that the b-axis tends to line up in the direction of maximum pressure or minimum tension, the a-axis in the direction of minimum pressure or maximum tension and the c-axis in that of the intermediate pressure. Kumazawa noted that the anisotropy in the Mid-Atlantic Ridge area was brought about by a prolonged stress field in which one of the principal axes is perpendicular to the ridge and the other along the ridge. If this view is correct, the upper layers near mid-oceanic ridges are in a stress field of which the tensional force is directed in the direction of spreading.

The development of preferred orientation of olivine crystals may also occur in the upper mantle of the island arcs. A note on this problem has already appeared (Sugimura and Uyeda, 1967). Our fundamental assumption is that the mantle convection current meets the continental mantle, which as a whole is immobile, and descends along the inclined seismic plane. It is then inferred that the velocity gradient would be concentrated near the seismic plane. The seismic plane dips about $30°-60°$ toward the continent, so that the general stress pattern in the mantle earthquake zone in island arc regions would be as shown by A in Fig.125. The principal axis of maximum pressure is almost horizontal and perpendicular to the trend of the island arc, while that of maximum tension or minimum pressure is in a vertical direction and the intermediate axis along the island arc, assuming the dip of the seismic plane to be $45°$. Under such a stress field, Kumazawa's theory predicts that the preferred orientation of olivine crystal would be as shown by B in Fig.125.

Such an alignment of olivine crystals would result in the P-wave velocity being abnormally high in the vertical direction but abnormally low in the horizontal direction compared with the normal mantle. Usually, the wave velocity in the vertical direction is difficult to assess observationally, so that the abnormally high velocity suggested here could well have escaped detection so far. Also, the anisotropy between the b- and c-axis is rather small and hard to confirm especially when the preferred orientation is not perfect

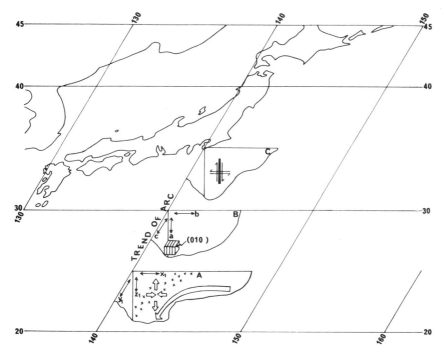

Fig.125. Diagrammatic sketch showing possible process under island arcs. A = major tectonic movement and principal axes x_1, y_1 and z_1 of regional stress; B = statistical orientation of the olivine crystallographic axes a, b and c as well as its cleavage plane (010); C = attitude and direction of motion on the focal slip plane as a double couple expression. (After Sugimura and Uyeda, 1967.)

and the mantle contains some isotropic crystals other than olivine. Thus, the anisotropy of seismic velocities expected from a mantle consisting of aligned olivine crystals should show the tendency that P-wave propagation in any horizontal direction is slower than the wave in the aggregate of randomly oriented crystals. This tendency may be a cause of what is revealed from observation of seismic waves in Japan (see p.24).

It will be shown that a logical consequence of the above idea leads to a possible explanation of observed radiation patterns of deep earthquake waves. As was noted in Fig.29, the stress pattern deduced from the initial motion of deep and intermediate earthquakes has remarkable regularity. As a source mechanism of earthquakes, the double couple mechanism has been established from observations of the first motions of S-waves. On the other hand, the dislocation theories have shown that a rupture due to slipping is the most probable form of movement in the earthquake foci and the source in this case can be constructed by the integration of the double couple force system generating the radiation pattern of waves. The slipping plane may be one of the two possible fault planes of this force system. The direction of maximum pressure in Fig.29 is nearly always perpendicular to the trend of the island arc. This is in agreement with the expected

h ≥ 100 km

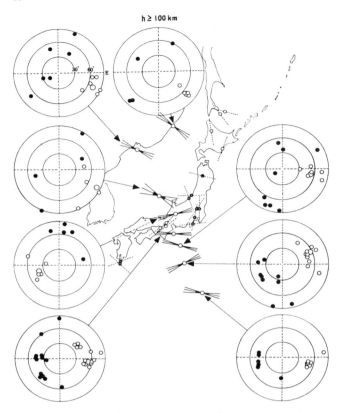

Fig.126. Geographical distribution of the directions of maximum pressure and tension, based on the radiation patterns of mantle earthquakes deeper than 100 km inclusive of shocks about 100 km deep. Stereographic directions are projected to upper hemispheres. An open circle and a closed circle in the stereographs indicate the directions of compression and extension, respectively. A single circle and a double circle in the map represent the deep and intermediate earthquakes, respectively. Three azimuths indicated by solid lines show the mean direction and its standard deviation of the compression projected to the earth's surface. One azimuth indicated by a dashed line shows the compression direction of a single earthquake. Note the exceptional attitude of the source mechanism shown in the stereograph for the inland area of central Japan. The fault planes in this area may be subparallel to those of the other areas, but the slip directions are reversed to those of the others. (After Ichikawa, 1966.)

direction of the maximum compressional axis when the mantle convection current is flowing towards the continent from the Pacific. However, in the vertical section, the direction of the maximum pressure is by no means horizontal but tends to be parallel with the deep seismic zone as shown in Fig.30, the slip-plane being either vertical or horizontal. If the current is flowing as we have assumed (Fig.125), the seismic zone would be the slip-plane and the principal stress axes would be horizontal and vertical when the inclination of the seismic plane is 45°. This discrepancy in the direction of stress axes has

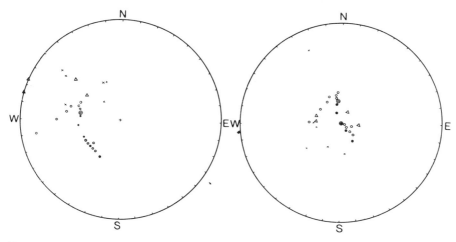

Fig.127. Stereographic projections showing the higher concentration of directions of minimum principal (compressional) stress to the dip-direction of the seismic plane (right) than the distribution of the same directions on the geographically spaced sphere (left). Both diagrams are projections to the lower hemisphere. All the directions for the mantle earthquakes deeper than 120 km except those beneath the land area of central Japan (see Fig.126) are indicated in the left diagram, in which the seismic plane not projected here shows different attitudes in different areas. These directions were rotated to the position in the right diagram where the seismic plane was imagined to be vertical and trend north–south. The centre of the right diagram means the dip-direction of the seismic plane. Symbols represent five different areas. (Data from Honda et al., 1957.)

been pointed out as a basis of the argument against the convection current hypothesis (e.g., Aki, 1966).

Fig.126, compiled by Ichikawa (1966), shows eight circles representing the stereographic projections upon the upper hemisphere of the directions of principal stress axis of deep and intermediate shocks. Circles are for different areas. Seven of these areas exclusive of the Honuriku area, the Sea of Japan side of central Honsyu, show the maximum pressure axes as subparallel to the steepest direction of the seismic plane. The degree of this parallelism can be expressed by Fig.127.

Average or smoothed radiation patterns were derived for sixteen selected areas in Japan, by determining a simple solution for a group of earthquakes (Aki, 1966). In this determination many shocks are regarded as if they were coming from a single source and the distributions of their first motions are superposed. Although Aki's data cover the period of only one and a half years from 1961 to 1963, the results essentially coincided with those obtained by H. Honda and his collaborators (see p.36) from the data on large earthquakes during 36 years. From the results of Aki's work, smoothed patterns in the areas around C–D and D–E (Fig.29) are reproduced in Fig.128. In the land area C–D in the Pacific coast of northeastern Honsyu, where the data are from deep earthquakes, the radiation pattern is similar to that of the area A–B in Fig 29. On the other hand, in the Pacific area off Honsyu, where the data are from shallower shocks, the axis of maximum

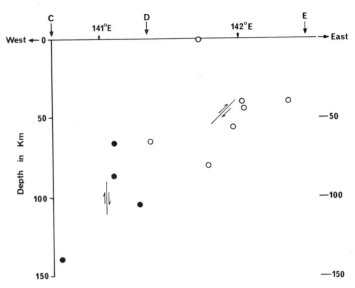

Fig.128. Cross-section across northeast Japan showing the estimated types of focal faults in the uppermost mantle. Positions C, D and E are plotted on Fig.29. Open circles are the hypocenters showing the thrust type in average stress pattern; closed circles are those of the vertical fault type in average stress pattern. (Data from Aki, 1966.)

tension is directed nearly vertically. This means that the slip displacement is a reverse dip-slip movement along a plane dipping either to the east or to the west. If we choose the fault plane dipping to the west, that would be the reverse faulting along sliding planes parallel to the seismic plane. Here it may be added that Stauder (1968) found that the forces are tensional for the shocks occurring outside the Aleutian Trench and compressional inside it. Aki's earthquakes in $D-E$ correspond to Stauder's latter shocks. As to the tensional features of outer shocks, bending of the lithosphere may be the cause as Isacks et al. (1968) proposed. To summarize, for relatively shallow earthquakes, the "focal" stress patterns can be regarded as being close to the "regional" one expected from the reverse faulting caused by the descending convective flow, whereas for deeper earthquakes, "focal" patterns would be produced at an angle of about $45°$ with the "regional" patterns.

Here we notice that the upper mantle may be anisotropic in the sense that olivine crystals have preferred orientations determined by the general stress field of Fig.125: the (010) planes of olivine may be statistically aligned in the vertical plane containing the strike of the island arc. Thus the upper mantle under the island arc has easily slipping planes in the vertical direction so that under a given stress field a part of the mantle would tend to slip preferentially along the vertical plane (C in Fig.125). This gives one possible solution for the problem which arises from the discrepancy between the direction of fault planes deduced from radiation patterns and the direction of maximum

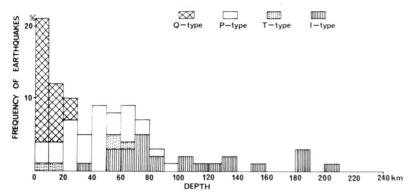

Fig.129. Histogram of frequency in types of earthquake mechanism in Japan. *Q*-type is the quadrant type, *P*-type is the compressional or thrust type, *T*-type is the extensional or normal-fault type, and *I*-type is the vertical-fault type. (Data from Aki, 1966.)

shear planes expected from the descending flow along the seismic plane. It gives also a logical basis for choosing the vertical plane from the two possible slip-planes deduced from the mechanism study.

There are many surface faults running parallel to the island arcs (Fig.31). Large strike-slip faults are known to be active along the island arcs. Allen (1962) suggested that these faults may have a causal relationship with the oceanic trenches. The problem can be approached from the directions of the mantle earthquake faults, since the larger surface faults would be of the deeper origin. Putting aside the question of the strike-slip movement, the formation of the slip plane may be the consequence of rearrangements of olivine crystals in the upper mantle due to the "regional" stress pattern (Fig.125).

A histogram (Fig.129) showing the frequency of occurrence of the different types of radiation patterns is obtained from the data of Aki (1966), where the *Q*-type indicates the strike-slip quadrant pattern, the *P*-type the reverse dip-slip, the *T*-type the normal dip-slip, and the *I*-type the pattern of either vertical or horizontal displacement. The *Q*-type earthquakes are restricted almost to the crust as is well known. From about 50 km depth to about 80 km depth, the *P*-type as well as *T*-type patterns vanish and the *I*-type ones begin to appear. Below these depths, the condition of plasticity would be satisfied, the oriented crystals following the "regional" stress pattern would increase, and thus the vertical slip planes would result. The authors' view that the transition boundary between the upper isotropic mantle and the lower anisotropic mantle underneath the island arcs would be from 50 km to 80 km depth is based on this observation.

Sinking lithosphere and deep earthquake zone

It is needless to say that the focal mechanism solutions of earthquakes are the most important source of information about the mechanical processes taking place in focal regions. Alternative to the interpretation outlined on p.196, Isacks et al. (1968) and Isacks and Molnar (1969) suggested a view which favours the concept of the lithosphere as a stress-guide (Elsasser, 1969).

As described already, a prominent feature of focal mechanism solution of intermediate and deep earthquakes is that their axes of maximum compression are parallel to the dip of the seismic plane. The above authors postulate that this feature can be explained if the stresses revealed by the focal solutions are not those acting between the sinking slab and the surrounding upper mantle, but are those acting *within* the sinking slab: i.e., the slab sinks downward into the asthenosphere, probably by its own weight, and the stresses within the slab are built up as it meets the resistance of the hard layer beneath the asthenosphere. This view was strengthened considerably by the discovery of evidence for down-dip extensional stress at the intermediate depth in several arc regions as shown in Fig.130. These regions are characterized by prominent gaps of seismicity at depth between 300—500 km. Apparently, while the leading part of the slab experiences compression parallel to the direction of sinking, the upper portion experiences a pull and the portion between the above two portions represents the level of no stress. This phenomenon has not been observed in the arcs around Japan as shown in Fig.130. The mechanism was suggested by Isacks and Molnar (1969) as shown in Fig.131 which is more or less self-explanatory. As one might notice, however, the view favoured by Isacks and his colleagues seems to be somewhat disturbed in the case of Japanese areas shown as

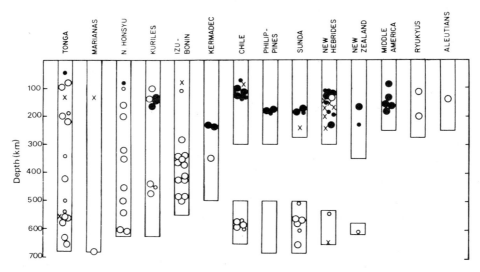

Fig.130. Down-dip stress type plotted as a function of depth for fourteen regions. A filled circle represents down-dip extension; an unfilled circle represents down-dip compression; and the crosses represent orientations that satisfy neither of the preceding conditions. Smaller symbols represent less reliable determinations. The enclosed rectangular areas approximately indicate the distribution of earthquakes as a function of depth by showing the maximum depths and the presence of gaps for the various zones. (After Isacks and Molnar, 1969.)

C in Fig.131. In the Northeast Honsyu Arc, the situations of slip movements were found as shown in Fig.128, which is representative but within a few exceptions. In other cross-sections, at the same depths as those of the foci plotted in the left half of Fig.128, there are some vertical faults having the westside down and the eastside up inferred from the radiation pattern (e.g., Ichikawa, 1966, fig.8), which can be interpreted also as the extensional stress along the seismic plane in terms of Isacks and others. Consequently, both the vertical faults with up slip movement and those with down slip movement of the ocean-side block exist together in the same level about 100 km deep in the northeast Honsyu Arc. This fact seems to us to indicate the primary importance of the parallelism of focal planes and the secondary importance of the sense of slip movements, and led us to consider that the easily sliding planes should be vertical and parallel to the island arc (Fig.125).

The thinning of the slab toward the tip is perhaps meant to represent the effect of gradual heating of the initially cold lithosphere. McKenzie (1969) made a heat conduction calculation of a sinking slab and showed the thinning very similar to those in Fig.131, on the assumption that ruptures causing earthquakes are only possible when the temperature is lower than a certain fraction of melting temperature, say 0.8. Of course, at the present stage, the spatial relationship of the so-called "Benioff zone" and the lithospheric plate has not been clarified from purely seismometric techniques, because the present accuracy of position determinations of hypocenters or of lithosphere is insufficient to allow this. Instead, it has been influenced by theoretical considerations. Namely, in the model that mantle earthquakes are due to the interaction of the sinking slab and the upper mantle, the seismic plane is placed at the upper boundary of the slab as shown in Fig.110 and 111, whereas in the model explained in this section, the seismic plane is assumed to be in the middle of the slab (Fig.131 and also Fig.24). The problem of

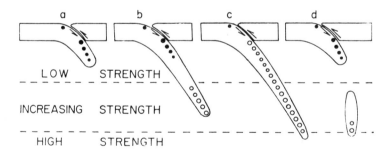

Fig.131. A diagram showing possible distribution of stresses in slabs of lithosphere that sink into the asthenosphere (*a*) and hit bottom (*b* and *c*). *d* represents the case where a piece of lithosphere has broken off. The symbols are the same as in Fig.130. In *b* and *d* gaps in the seismicity would be expected. Also shown is the underthrusting and the extensional stresses near the upper surface of the slabs due to the bending of the slab beneath the trenches. These features are inferred from the mechanisms of shallow earthquakes. The lower boundary where the slabs hit bottom might correspond to the transition or discontinuity near 650–700 km. (After Isacks and Molnar, 1969.)

clarifying which of the above two types of models is closer to the actual case would be extremely important, because it would be a decisive factor in the consideration of the thermal processes under island arcs, including the identification of the force which is directly responsible to the movements of plates: whether it is the thermal convection current under the plates or the weight of sinking plates themselves. Recently, Carr et al. (1972) showed that the deep and intermediate foci under Japan are concentrated on two planes that may bound the undergoing slab, the concentration being clearer on the upper surface of the slab than on the lower. This fact might be a negative evidence against the latter type model that these shocks would be generated within the slab. The shearing stress along the boundary surfaces of the slab must generate these earthquakes as we thought before (Sugimura and Uyeda, 1967).

THE ISLAND-ARC TYPE OROGENY

There is no doubt that the Mizuho orogeny in Japan includes the present island arc formation and its activity as noted in Chapter 2. It is also conceivable that during the whole period of the Mizuho orogeny east Japan at least has shown the island arc features: the characteristic topography inclusive of trenches, active volcanoes, and even deep earthquakes. However, there are some questions to consider as to whether the orogeny in the mesotectonic age showed the exactly similar island arc features or not. For example, Matsumoto (1967) distinguished the island arcs from the circum-Pacific orogenic belts of the mesotectonic age, based on reasons which include the fact that the latter lacks the marginal seas behind, although they resemble each other in many respects. His opinion appears to represent the widely held ideas of many Japanese geologists. That the mesotectonic circum-Pacific orogenies show basaltic activity in their outer belts is also one of the differences from the island arcs. We could not simply put the two orogenies in the same class, before it could be made clear that the difference in type is only one of appearance or only of secondary importance.

However, the concept of paired metamorphic belts, originally proposed by Miyashiro (1961a) from the study of the Late Mesozoic orogeny in Japan and other circum-Pacific areas, has been developed along with the view that the modern island arc features are nothing but a present manifestation of the paired orogeny. Miyashiro (1961a) pointed out that the Mesozoic metamorphic belt in the circum-Pacific region consists of outer and inner belts having contrasting natures in metamorphism and in associated magmatism. He further pointed out that the outer metamorphic belt, characterized by high p/t type metamorphism, lies always on the oceanic side area which has no sialic basement. Miyashiro (1959, 1961b) noted that the two types of regional metamorphism, i.e., the high p/t type and low p/t type may presently take place under the trench and inner belts of island arc, respectively. Later, Takeuchi and Uyeda (1965) drew attention to the remarkable coincidence of the $p-t$ conditions under the trench and inner belts estimated from heat flow data (Uyeda and Horai, 1964) with the $p-t$ conditions for the two types of metamorphism estimated from mineralogical grounds (Fig.107). Miyashiro (1967) and the present authors in this monograph proposed a model for the origin of the paired zone putting the prime importance on the mantle convection sinking along the deep-seismic plane beneath the island arc system.

Thus, the authors argue that the orogenic movement of the mesotectonic age is also essentially one of the island-arc type orogenies. Miyashiro (1959) called this type of orogeny the Japan-type orogeny, and Matsuda and Uyeda (1971) called it the Pacific-type orogeny and discussed its possible mechanism. Most materials developed here have been published by Matsuda and Uyeda, but the argument put forth here differs slightly from them.

The Pacific-type orogeny

The Pacific-type orogeny may appear to be only a local variety of the classical type of orogeny established in the European and American continents (e.g., Kay, 1951; Aubouin, 1965). But, in fact, there seem to be many fundamental differences between them. For example, in the generally accepted but recently criticized pattern of orogenic evolution, the "geosynclinal phase" with mafic volcanism is followed by an "orogenic phase" with granitic magmatism in the same belt. In an orogenic belt with the paired belts, however, the geosynclinal environments and the activity of granitic magmas prevail almost simultaneously in the juxtaposed outer and inner belts. The formation of these paired belts occurs at the margin of continental lithosphere, where the consumption of oceanic lithosphere takes place. On the contrary, the classic model of the orogenic belt has been studied in the boundary areas between two continental lithospheres pushing against each

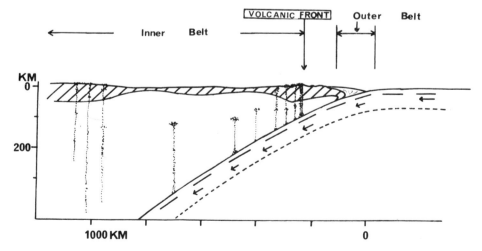

Fig.132. A schematic profile of the Pacific-type orogenic belt. (After Matsuda and Uyeda, 1970.)

other. The latter type may be called the Alpine–Himalayan-type orogeny as opposed to the Pacific-type.

The framework of the Pacific-type orogenic belt is asymmetric and paired, consisting of the Outer and Inner Belts as schematically shown in Fig.132. The Outer Belt is the place of regional subsidence, sedimentation and the high p/t type metamorphism, as suggested by the negative isostatic anomaly of gravity and low heat flow in active island arcs. On the other hand, the Inner Belt is characterized by various types of magmatism and high heat flow. The features of the Inner Belt extend more than 1,000 km from the volcanic front to the interior of the continent. The Outer and Inner Belts are separated, as a rule, by a non-volcanic geanticlinal ridge. Thus, the narrow Outer Belt on the oceanic side, the wide Inner Belt on the continental side, and the median geanticlinal belt form the basic framework of the Pacific-type orogenic belt. A Late Cenozoic example of paired belts in northeast Japan has already been shown on p.108.

Late Mesozoic–Early Cenozoic paired belts in southwest Japan

Earlier the Hirosima belt (see p.93) of acidic igneous activity was assigned to be a part of the Inner Belt, which extends to the large area of the Asiatic continent (Fig.133). Welded rhyolitic pyroclastic flow deposits similar to those in the Hirosima belt were found from submarine highs, the Yamato Bank and other banks in the southwestern Sea of Japan (Sato and Ono, 1964). Some parts of the southwestern Sea of Japan floor should be the fragments of the Inner Belt. As pointed out by many authors (e.g., Ichikawa et al., 1968; Matsumoto, 1968), it is noteworthy that these large scale intrusions and extrusions of granitic or rhyolitic magmas were not preceded immediately by geosynclinal subsidence, but by localized marine and lacustrine basins (Research Group for Late Mesozoic Igneous Activity of southwest Japan, 1967; Ichikawa et al., 1968). Before the Late Cretaceous when this igneous activity began, most of the area had been denudated since the Triassic when the Late Paleozoic to the earliest Mesozoic geosyncline turned into land. It may represent one of the characteristic features of the Pacific-type orogeny. Matsuda and Uyeda (1971) suggested that the Inner Belt included the Ryoke belt along its oceanic margin. Another alternative is that the Ryoke low p/t metamorphic belt shows the earlier and deeper facies of the Inner Belt than the Hiroshima belt, and plays the counter part of the Sanbagawa high p/t metamorphic belt, whereas only the Hiroshima belt may be correlated in age with the low-grade metamorphism in the outer Simanto belt. In both cases the acidic igneous activity took place over a very large areal extent and reminds us of the extensive high heat flow region in the present Inner Belt.

Various field evidences indicate that the Ryoke, Sanbagawa and Titibu belts were gradually rising in the latest Cretaceous and Paleogene periods as a geanticlinal ridge separating the Inner Belt from the Outer Belt. These belts are assigned to be the Median Belt, from which Matsuda and Uyeda (1971) excluded the Ryoke belt. In any way, the geanticlinal Median Belt seems to have existed just like the Late Cenozoic Median Belt.

Fig.133. Cretaceous–Early Tertiary granites and the contemporaneous outer sedimentary terrain (Simanto belt) in the Japanese islands and vicinity. (After the Geological Survey of Manchuria, 1940; Geological Survey of Korea and Geological Society of Korea, 1956; Ministry of Geology, U.S.S.R., 1965; Geological Survey of Japan, 1968; arranged by Matsuda and Uyeda, 1971.)

The rocks of the Simanto belt are isoclinally folded with the axial planes dipping north with relatively high angles (e.g., Harata, 1964; Hashimoto, 1966; Kimura, 1967; Yamada et al., 1969) and are regionally metamorphosed up to the degree of the greenschist facies through pumpellyite–prehnite metagraywacke facies (Matsuda and Kuriyagawa, 1965). This belt appears to be the late mesotectonic equivalent of the present Outer or Trench Belt.

Thus, the deeper geological features of the Pacific-type orogeny may be exemplified by southwest Japan in the latest Cretaceous to the Paleogene, the shallower features being exemplified by the northeast Japan three belts (see p.108). The paired metamorphism of the Ryoke and Sanbagawa belts may be the further deeper facies of this type of orogeny.

Possible mechanism of the Pacific-type orogeny and the origin of marginal seas

As illustrated in the preceding sections, major features of mesotectonic and neotectonic orogenies in the Japanese area are well described by pairs of contemporaneous Outer and Inner Belts having contrasting characteristics. A spatial separation of the geosynclinal environment and the igneous activity is the most important feature of the Pacific-type orogeny. The evolutional pattern of the classic type of eugeosyncline, starting with the initial magmatism of the geosynclinal stage, followed by the granitic magmatism and orogenic tectonism of subsequent stage, and ending with the final volcanism of post-orogenic stage, is currently criticized in Europe, and at least is applicable to neither the Inner Belt nor the Outer Belt of the Pacific-type orogeny.

Taking the various characteristic features of the Pacific-type orogeny into consideration, we propose a model for its process as shown in Fig.110. In the Outer Belt, the sea-floor is dragged down to form the trench inside of which the geosynclinal sedimentation proceeds. Low heat flow and shallow earthquakes are expected there. As was explained on p.167, we assume that the deep earthquakes along the inclined seismic plane are caused by the shear between the sliding lithosphere and the mantle above it, and the same shear causes generation of heat along the seismic plane to give rise to the partial melting. The molten material will be the parent of the magma in the Inner Belt. This model was examined recently by Hasebe et al. (1970) through numerical experiments (see p.168). Their results showed that a heat generation some five times as high as the mean heat flow (1.5 H.F.U.) at the upper surface of the slab, and an effective heat conduction ten times as large as the normal phonon conduction are required to account for the observed high heat flow in the Sea of Japan area. As was pointed, one interesting aspect of these results is that the amount of material flux needed to provide an effective vertical thermal conductivity ten times greater than the normal one would be some 0.3 cm/year, so that the total thickness of the material that has ascended during 10^8 years would become 300 km. A part of this material may be water, but a considerable part of this amount should produce the lithosphere and this is not inconsistent with the idea that the Sea of Japan was formed by an opening (e.g., Murauchi, 1966; Karig, 1969): The present Japanese Islands were originally the eastern marginal parts of the Asiatic continent, but after the commencement of the process envisaged here, magma has continued to rise so that the Japanese Islands had to drift southeastward to provide room for the magma (Fig.115). Present considerations, thus, call for a revival of the seaward drift of the Japanese Islands (Tokuda, 1926; Terada, 1934; Kobayashi, 1956). If the model is applicable to other active island arcs, it may be postulated that other marginal seas are also expanding by intrusions of magma from the seismic plane. In the plate tectonics or the new global tectonics, trench—arc systems are considered to be the area where plates are consumed. Matsuda and Uyeda (1971), however, state that at least part of the descending lithosphere would rise again to form the upper mantle of the marginal seas. But this idea contradicts the model of the zone partial melting, which implies only a small amount of mass ascent. Another question open to discussion is the difference between the granitic or rhyolitic magmas in the mesotectonic Inner Belt and the basic or ultrabasic magmas that are supposed to have produced the oceanic lithosphere of the Sea of Japan floor.

After all, it would be hardly conceivable that the granitic rocks of the Late Mesozoic and the Paleogene are the products of a partially molten zone above the seismic plane, because the source material of granitic rocks is thought to be different from that of volcanic rocks and probably originated in the continental crust. However, granitic intrusion may signify high terrestrial heat flow in the continental crust. If it does so, the extensive distribution of the Late Mesozoic and Paleogene granitic intrusion and rhyolitic extrusion shows one of the important relics of the island arc type orogeny and may reflect an extensive consumption of the Pacific plate in that time.

ISLAND ARC TECTONICS AND OCEANIC RIDGE TECTONICS

World encircling systems of mid-oceanic ridges and rift valleys (Fig.1) have been the subject of intensive investigations in the last decade. The features of mid-oceanic ridges are characterized by linear crestal continuation of high heat flow, shallow earthquakes, volcanic islands, and magnetic anomalies, let alone the topographical highs. Many of these systems show steep central rifts. Strike-slip faults develop across the crest of mid-oceanic ridges at many places. Focal mechanism study by Sykes (1967) showed that the displacements accompanying the earthquakes on these faults are of the transform fault type (Wilson, 1965). Crustal structures are oceanic in that no granitic crust is developed. But, seismic Pn velocity as well as gravity measurements over the Mid-Atlantic Ridge (Talwani et al., 1965) indicated that the upper mantle is probably quite anomalous in the sense that both Pn and density are small. It is now generally accepted that mid-oceanic ridges and continental rift valleys (Girdler, 1963) may be the surface manifestations of a common process in the upper mantle which might be the rising mantle convection current. According to the classic and newly revived hypothesis of continental drift, the rising convection current now forming the Mid-Atlantic Ridge was originally under the super-continent that is now separated into two by the Atlantic Ocean (Holmes, 1929; Runcorn, 1962; Wilson, 1963). General occurrence of oceanic ridges in the median position between continents can be interpreted likewise (Menard, 1965). The hypothesis of ocean-floor spreading (Dietz, 1961; Hess, 1962) now seems to be well founded by the celebrated matches between the magnetic lineation patterns and the reversals of the geomagnetic field in the geological past (Vine and Matthews, 1963; Vine and Wilson, 1965; Cox et al., 1967). According to the recent results of JOIDES (Joint Oceanographic Institutions Deep Earth Sampling) drilling, the ages of the basement rocks of ocean floors have been found to be in good agreement with the ages expected from the spreading hypothesis (Maxwell et al., 1970). In some places, such as the Gulf of Aden and the Red Sea, or the Gulf of California, it is considered that the mid-oceanic ridges come into the continents and that the splitting of continents is now in progress (Girdler, 1962). Concerning these areas, Wilson (1965) proposes a modified view based on the concept of the transform fault. In contrast to the island arcs, these systems have neither trenches nor deep earthquakes.

The authors agree with the idea that the island arcs and orogenic zones and the mid-oceanic ridges and rifts are the two main features of basic importance and the essential difference between them is that the former originates from the descending convection current whereas the latter is from the rising current. Therefore, even though some phenomena are common to both structures in appearance, the mechanism of producing them is probably considerably different. The attempts to establish a unified tectonic view by, for example, Morgan (1968), Le Pichon, (1968), and Isacks et al. (1968) are much favoured by the present authors, who wish to make a similar attempt, putting more emphasis on the thermal aspects. Recently J. F. Dewey and his colleagues (Dewey, 1969a, b; Bird and Dewey, 1970; Dewey and Bird, 1970; Dewey and Horsfield, 1970) have been developing a remarkably unified theory on orogeny based on the plate tectonics.

According to Kuno (1964), some oceanic basalts have compositions closer to peridotite (closer to the bottom left corner on Fig.120) than the island arc basalts. Kuno interprets this observation as due to the basic difference in the mechanism of magma production in oceanic ridges and island arcs (Fig.116). In the case of oceanic basalt, melting is supposed to be induced by pressure release in the hot ascending column of the mantle material. Therefore the melting tends to be more progressive than in the case of island arc basalt production which is, in the authors' opinion, due only to the heat generated by the shearing process in the boundary layer between the relatively cold descending current and the immobile continental mantle. The amount of thermal energy available for the melting is likely to be different in the two cases.

McBirney and Gass (1967) demonstrated successfully the zonal arrangement in chemical composition of primary magmas on both sides of mid-oceanic ridges. The substantial point of this arrangement, that the silica-rich magma is close to the ridge and the alkaline magma is far away from the ridge, is just the same as for the island arc primary magmas in regard to the trench (see p.58). Both arrangements seem to show that the chemical composition of the liquid is controlled not by that of the original rocks from which the magma derived, but by the pressure under which the magma was produced. The difference between the two arrangements is caused by the difference in the configuration of isotherms, which determines the distribution of pressure in the magma generation zone. In mid-oceanic regions the zonal arrangement of primary magmas occurs on both sides of the ridge with a mirror symmetry, whereas in island arc regions it appears only on one side of the trench with a distinct front. The distance concerned is also different. The width of the magma provinces around mid-oceanic ridges is larger than those of the island arc volcanic belts. This difference suggests that the slope of isotherms around the ridges is gentler than that beneath the island arcs.

A basic difference is likely to exist in the anisotropic nature of the upper mantle in these two types of tectonic areas. Since the dominant stress fields are different, the directions of preferred orientation of olivine crystals are supposed to be different, as already explained (see p.192). The proposed preferred directions of olivine crystals in the two cases are listed in Table XI.

In the theory of sea-floor spreading, the magnetic lineations or isochrons are parallel to

TABLE XI

Proposed preferred directions of olivine crystals.

	Island arc region	Mid-oceanic ridge area
Along the axis	c-axis	b-axis
Across the axis	b-axis	a-axis
Vertical direction	a-axis	c-axis

the source of the new oceanic bottom, i.e., the mid-oceanic ridges. In the northwestern Pacific off Japan the trend of the lineations is northeast—southwest (Fig.46). If the source of these lineations is located as far as the East Pacific Rise, these lineations must have migrated over the entire Pacific Ocean and are about to sink down underneath the Kurile Trench. This view seems to fit the overall picture of the hypothesis. But this simple-minded expectation may not be supported by field evidence. In fact, there is a possibility that these lineations may be linked with those in the eastern Pacific in a rather striking manner: Peter (1966) found that the north—south trending lineations in the eastern Pacific bend sharply on approaching the Aleutian Trench. Hayes and Heirtzler (1968) and Erickson and Grim (1969) further traced the westward continuation of the lineations to almost $170°$ E longitude as shown in Fig.134. Although the east-west trending lineations south of the Aleutians appear to be interrupted by the Emperor Seamounts Ridge, they may be connected with the lineations in the northwestern Pacific. If that is the case, the source of the spreading in the northwestern Pacific area must be in the northwestern side of the lineations and the direction of spreading should be the reverse of the case that the lineations were formed in the East Pacific Rise area. Further, in this case, it would be difficult to regard the trench as the sink of the spreading floor. This would put the whole picture of island arc tectonics presented in the present monograph into an apparent difficulty. The same difficulty is already demonstrated in the case of the Aleutian arc and trench system. Pitman and Hayes (1968) interpret the situations there in terms of the notion of migrating ridge. The former sources of the lineations in the Gulf of Alaska and south of the Aleutians were forced to migrate northeastward and northward, respectively, and are now probably buried underneath Alaska and the Bering Sea. The concept of a migrating ridge under a trench (McKenzie and Morgan, 1969) has recently been demon-strated for the west coast of North America by Atwater (1970). If the lineations off Japan are linked with those off the Aleutians, the same situation may hold for the arcs of Japan. The Baikal Rift might be the present manifestation of the migrated ridge. But, the formation of the sea-floor as the hypothetical ridge migrated, if it really happened, must have taken place before the present island arc activity started. In this respect, one should note that the lineations off Japan are supposed to be quite old. In fact, the Peter's guide anomaly is the anomaly no.31 in Heirtzler et al. (1968) which is considered to be about $80·10^6$ years old. These lineation patterns should not necessarily represent the pattern of

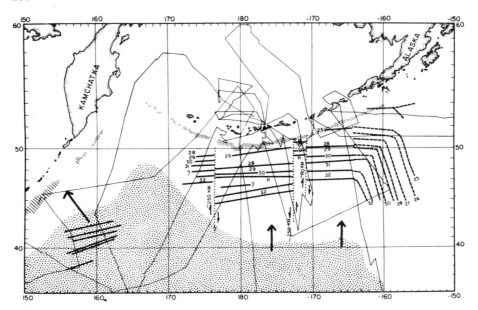

Fig.134. Map of magnetic lineations (bold lines) and lineation offsets (dotted lines) in the area south of the Aleutians. The Aleutian and Kurile trenches are represented by the hatched areas. The areal extent of the opaque sediment horizon of Ewing et al. (1968) is given by the stippled area. Tracks of "Vema" and "R. D. Conrad" are also shown. (After Hayes and Heirtzler, 1968.) Thick arrows indicate the direction of the present relative movement inferred by the present authors.

the *present* motion. It is likely that the patterns were formed long ago and that the present arc and trench system is formed by the later relative motion of the oceanic and continental plates. The ancient source of the lineations which might reside under the other side of the arcs may have become inactive already and have ceased to be a source.

As to the continuity of lineations of the east and west Pacific, Hayes and Heirtzler (1968) suggested that the lineations off Japan may be much older than those in the east Pacific and the south of the Aleutians, so that they are not linked. The existence of *Horizon A* in the sediment layer (Ewing and Ewing, 1967) in the western Pacific seems to support their view. In such a case, the direction of the lineations have little to do with the present movement of the mantle. At any rate, the age of the floor of the northwest Pacific off Japan is a mystery at this moment. The estimated directions of the present flow relative to the arcs are as shown in Fig.134 by thick arrows, and these directions do not conform neither of the simple flow patterns suggested by the ancient lineations. The directions of the movement of the Pacific Ocean floor in recent times indicated by the arrows are supported by the paleomagnetic data on seamounts off Japan (Uyeda and Richards, 1966; Vacquier and Uyeda, 1967). These data suggest that the seamounts were nearly on the magnetic equator when magnetized and have migrated to the present latitudes. Molluscan fossils of nerineids of the Cenomanian to the Senonian Epoch were found in limestone from one of these seamounts (Tsuchi and Kagami, 1967). The nerineids as well as other organisms found in this limestone belong to the fauna and flora of the warmer water.

The time variations in the pattern of motions may be important in interpreting the observed magnetic isochrons of the ocean floor and the system of fracture zones. Le Pichon (1968) developed an elaborate geometrical model of the earth's surface from the observations of magnetic lineations and fracture zones. In this argument, it was inferred that motion of the north Pacific indicated by the fracture zones off the North American continent is ancient and that the present motion is a rotation characterized by the spreading from the south Pacific. This would agree with the above view that the possible continuation of the East Pacific Rise which played the role of the source of lineations in the north Pacific died or at least weakened at present, by overriding of the American continent.

If the lineations off Japan are not linked with those off the Aleutians, it may be inferred that the lineations we observe off Japan were formed in Late Paleozoic to Mesozoic age by some spreading centre that does not exist any more. The hypothetical Darwin Rise may be a possibility of such a spreading centre. But the reality of Darwin Rise has been seriously questioned by the JOIDES work (Deep Sea Drilling Project, 1969). At any rate, the changes of trend of orogenic belts represented by the migrations of volcanic fronts (Fig.75, 76) may provide important key information regarding the history of the evolution of the Pacific Ocean floor as well as that of the circum-Pacific island arcs.

REFERENCES

Aki, K., 1961. Crustal structure in Japan from the phase velocity of Rayleigh waves, 1. Use of the network of seismological stations operated by the Japan Meteorological Agency. *Bull. Earthquake Res. Inst.,* 39: 255–283.

Aki, K., 1966. Earthquake generating stress in Japan for the years 1961 to 1963 obtained by smoothing the first motion radiation patterns. *Bull. Earthquake Res. Inst.,* 44: 447–471.

Aki, K., 1967. Scaling law of seismic spectrum. *J. Geophys. Res.,* 72: 1217–1231.

Aki, K and Kaminuma, K., 1963. Phase velocity of Love waves in Japan, 1. *Bull. Earthquake Res. Inst.,* 41: 243–259.

Akimoto, S. and Fujisawa, H., 1965. Demonstration of the electrical conductivity jump produced by the olivine–spinel transition. *J. Geophys. Res.,* 70: 443–449.

Akimoto, S. and Fujisawa, H., 1968. Olivine–spinel solid solution equilibria in the system $Mg_2 SiO_4 - Fe_2 SiO_4$. *J. Geophys. Res.,* 73: 1467–1479.

Allen, C. R., 1962. Circum-Pacific faulting in the Philippines–Taiwan region. *J. Geophys. Res.,* 67: 4795–4812.

Anderson, D. L., 1966. Recent evidence concerning the structure and composition of the earth's mantle. In: L. H. Ahrens, F. Press, K. Rankama and S. K. Runcorn (Editors), *Physics and Chemistry of the Earth.* Pergamon, London, 6: 1–131.

Andreyeva, I. B. and Udintsev, G. B., 1958. Bottom structure of the Sea of Japan, from the "Vityaz" expedition data. *Izv. Akad. Nauk S.S.S.R., Geol. Ser.,* 10.

Aramaki, Sh., 1965. Mode of emplacement of acid igneous complex (Kumano Acidic Rocks) in southeastern Kii Peninsula. *J. Geol. Soc. Japan,* 71: 525–540. (In Japanese with English abstract.)

Aramaki, Sh. and Hada, Sh., 1965. Geology of the central and southern parts of the acid igneous complex (Kumano Acidic Rocks) in southeastern Kii Peninsula. *J. Geol. Soc. Japan,* 71: 494–512. (In Japanese with English abstract.)

Argand, E., 1916. Sur l'arc des Alpes occidentales. *Eclogae Geol. Helv.,* 14: 145–191.

Atwater, T., 1970. Implications of plate tectonics for the Cenozoic tectonic evolution of western North America. *Geol. Soc. Am., Bull.,* 81: 3513–3536.

Aubouin, J., 1965. *Geosynclines – Developments in Geotectonics, 1.* Elsevier, Amsterdam, 335 pp.

Aumento, F., 1969. Geology of the Mid-Atlantic Ridge at 45°N. Paper presented for the discussion on *Petrology of Igneous and Metamorphic Rocks from the Ocean Floor,* 13 Nov. 1969, London. (Unpublished.)

Barazangi, M. and Dorman, J., 1969. World seismicity maps compiled from ESSA, Coast and Geodetic Survey, epicenter data., 1961–1967. *Bull. Seismol. Soc. Am.,* 59: 369–380.

Bénard, H., 1901. Les tourbillons cellulaires dans une nappe liquide transportant de la chaleur par convection en regime permanent. *Ann. Chim. Phys.,* 23: 62–144.

Benioff, H., 1949. The fault origin of oceanic deeps. *Bull. Geol. Soc. Am.,* 60: 1837–1866.

Benioff, H., 1954. Orogenesis and deep crustal structure. Additional evidence from seismology. *Bull. Geol. Soc. Am.,* 65: 385–400.

Benioff, H., 1955. Seismic evidence for crustal structure and tectonic activity. *Geol. Soc. Am., Spec. Papers,* 62: 61–73.

Birch, F. and Clark, H., 1940. Thermal conductivity of rocks and its dependence upon temperature and composition. *Am. J. Sci.,* 238: 529–559; and 611–635.

Bird, J. M. and Dewey, J. F., 1970. Lithosphere plate-continental margin tectonics and the evolution of the Appalachian orogen. *Geol. Soc. Am., Bull.,* 81: 1031–1061.

Block, M. J., 1956. Surface tension as the cause of Bénard cells and surface deformation in a liquid film. *Nature,* 178: 650–651.

Bodvarsson, G., 1955. Terrestrial heat balance in Iceland. *Timarit Verkfraeoingafelags Islands,* 39: 1–8.

Bogdanoff, A., 1963. Sur le terme "étage structural". *Rev. Géogr. Phys. Géol. Dyn.*, 5: 245–253.

Bowen, N. L. and Tuttle, O. F., 1949. The system $MgO-SiO_2-H_2O$. *Bull. Geol. Soc. Am.*, 60: 439–460.

Boyd, F. R., 1964. Geological aspects of high-pressure research. *Science*, 145: 13–20.

Boyd, F. R. and England, J. L., 1963. Effect of pressure on the melting of diopside $CaMgSi_2O_6$ and albite $NaAlSi_3O_8$ in the range up to 50 Kb. *J. Geophys. Res.*, 68: 311–324.

Bracey, D. R., 1963. Marine magnetic profiles in the Pacific Ocean, 1961–1962. *Marine Sci. Dept., U.S. Naval Oceanogr. Office, Inf. Manuscr. Rept.*, M-4-63. (Unpublished.)

Bracey, D. R., 1966. Geomagnetic measurements in the Pacific Ocean aboard U.S.N.S. Charles H. Davis (AGOR-5), 1964. *Hydrogr. Surv. Dept., U.S. Naval Oceanogr. Office, Inf. Rept.*, M-4-66. (Unpublished.)

Brune, J. M., 1968. Seismic moment, seismicity, and rate of slip along major fault zones. *J. Geophys. Res.*, 73: 777–784.

Bullard, E. C., Maxwell, A. E. and Revelle, R., 1956. Heat flow through the deep sea-floor. In: *Advances in Geophysics*. Academic Press, New York, N.Y., 3: 153–181.

Carr, M. J., Stoiber, R. E. and Drake, C. L., 1972. Seismicity and upper mantle structure under the Japanese arcs. *Geol. Soc. Am., Bull.* (In press.)

Carter, N. L. and AvéLallemant, M., 1970. High temperature flow of dunite and peridotite. *Geol. Soc. Am., Bull.*, 81: 2181–2202.

Cenozoic Research Group of Southwest Japan, 1960. An outline of the Cenozoic history of southwest Japan. *Earth Sci.*, 50/51: 56–65. (In Japanese with English abstract.)

Chandrasekhar, S., 1961. *Hydrodynamic and Hydromagnetic Stability*. Clarendon, Oxford.

Chinzei, K., 1966. Younger Tertiary geology of the Mabechi River valley, northeast Honshu, Japan. *J. Fac. Sci., Univ. Tokyo*, 16: 161–208.

Chinzei, K., 1967. An attempt for absolute chronology of Neogene in Japan by biostratigraphic correlation. *J. Geol. Soc. Japan*, 73: 220–221. (In Japanese.)

Chinzei, K., 1968. Evolution of the Japanese Islands. In: A. Miyashiro and S. Uyeda (Editors), *Cosmic Sciences and Earth Sciences*. Meiji Tosho, Tokyo, pp.208–235. (In Japanese.)

Christensen, M. N., 1965. Late Cenozoic deformation in the central Coast Ranges of California. *Geol. Soc. Am., Bull.*, 76: 1105–1124.

Clark Jr., S. P., 1957. Radiative transfer in the earth's mantle. *Trans. Am. Geophys. Union*, 38: 931–938.

Clark Jr., S. P. and Ringwood, A. E., 1964. Density distribution and composition of the mantle. *Rev. Geophys.*, 2: 35–88.

Coats, R. R., 1962. Magma type and crustal structure in the Aleutian Arc. In: G. A. MacDonald and H. Kuno (Editors), *The Crust of the Pacific Basin – Am. Geophys. Union, Geophys. Monogr.*, 6: 92–109.

Coode, A. H. and Tozer, D. C., 1965. Low-velocity layer as a source of the anomalous vertical component of geomagnetic variations near the coast. *Nature*, 205: 164–165.

Cox, A., Dalrymple, G. B. and Doell, R. R., 1967. Reversals of the earth's magnetic field. *Sci. Am.*, 216: 44–54.

Dambara, T., 1968. Vertical movements of Japan during the past sixty years, 4. Chubu district. *J. Geod. Soc. Japan*, 13: 66–74. (In Japanese with English abstract.)

Dambara, T. and Hirobe, M., 1964. Vertical movements of the earth's crust in the southern part of the Kanto district. *J. Geod. Soc. Japan*, 10: 146–153.

Davies, G. L. and Brune, J. N., 1971. Regional and global fault slip rates from seismicity. *Nature*, 229: 101–107.

Davis, B. T. C. and England, J. L., 1964. The melting of forsterite up to 50 Kb. *J. Geophys. Res.*, 69: 1113–1116.

Deep Sea Drilling Project, 1969. Leg 7. *Geotimes*, 14: 12–14.

Den, N., Ludwig, W. J., Murauchi, S., Ewing, J. I., Hotta, H., Edgar, T., Yoshii, T., Asanuma, T., Hagiwara, K., Sato, T. and Ando, S., 1969. Seismic refraction measurements in the northwest Pacific Basin. *J. Geophys. Res.*, 74: 1421–1434.

Dewey, J. F., 1969a. Continental margins: a model for conversion of Atlantic type to Andean type. *Earth Planet. Sci. Lett.*, 6: 189–197.

Dewey, J. F., 1969b. Evolution of Appalachian/Caledonian orogen. *Nature*, 222: 124–129.

Dewey, J. F. and Bird, J. M., 1970. Mountain belts and the new global tectonics. *J. Geophys. Res.*, 75: 2625–2647.

Dewey, J. F. and Horsfield, B., 1970. Plate tectonics, orogeny and continental growth. *Nature*, 225: 521–525.

Dickinson, W. R. and Hatherton, T., 1967. Andesitic volcanism and seismicity around the Pacific. *Science*, 157: 801–803.

Dietz, R. S., 1961. Continent and ocean basin evolution by spreading of the sea floor. *Nature*, 200: 1085.

Dietz, R. S., 1962. Ocean-basin evolution by sea-floor spreading. In: S. K. Runcorn (Editor), *Continental Drift*. Academic Press, New York, N.Y.

Diment, W. H., Ortiz, F. O., Silva, L. R. and Ruiz, C. F., 1965. Terrestrial heat flow at two localities near Vallenar, Chile. *Trans. Am. Geophys. Union*, 46: 175.

Duda, S. J., 1965. Secular seismic energy release in the circum-Pacific belt. *Tectonophysics*, 2: 409–452.

Elsasser, W. M., 1963. Early history of the earth. In: *Earth Science and Meteoritics*. North-Holland, Amsterdam, pp.1–30.

Elsasser, W. M., 1966. Thermal structure of the upper mantle and convection. In: P. M. Hurley (Editor), *Advances in Earth Science*. M.I.T. Press, Cambridge, Mass., pp.461–502.

Elsasser, W. M., 1969. Convection and stress propagation in the upper mantle. In: S. K. Runcorn (Editor), *The Application of Modern Physics to the Earth and Planetary Interiors*. Interscience, New York, N.Y., pp.223–246.

Elvers, D. J., Mathewson, C. C., Kohler, R. E. and Moses, R. L., 1967. Systematic ocean surveys by U.S.C. and G.S.S. "Pioneer", 1961–1963. *ESSA Operational Data Rept., C and GS DR-1*, 19 pp.

Emiliani, C., 1964. Paleotemperature analysis of the Caribbean cores A254-BR-C and CP-28. *Geol. Soc. Am., Bull.*, 75: 129–144.

Erickson, B. H. and Grim, P. J., 1969. Profiles of magnetic anomalies south of the Aleutian Island Arc. *Geol. Soc. Am., Bull.*, 80: 1387–1390.

Ewing, J. and Ewing, M., 1967. Sediment distribution on the mid-ocean ridges with respect to spreading of the sea-floor. *Science*, 156: 1590–1592.

Ewing, J., Ewing, M., Aitken, T. and Ludwig, W. J., 1968. North Pacific sediment layers measured by seismic profiling. In: L. Knopoff, C. Drake and P. Hart (Editors), *The Crust and Upper Mantle of the Pacific Area — Am. Geophys. Union, Geophys. Monogr.*, 12: 147–173.

Ewing, M. and Heezen, B., 1955. Puerto Rico Trench and geophysical data. *Geol. Soc. Am., Spec. Papers*, 62: 255–267.

Ewing, M., Worzel, J. L. and Shurbet, G. L., 1957. Gravity observation at sea in U.S. submarines Barracuda, Tusk, Congor, Argonaut and Medregal. *Verh. Koninkl. Ned. Geol. Mijnbouwk. Genoot., Geol. Ser.*, 18: 49–115.

Fedotov, S. A., Averjanova, V. N., Bagdasarova, A. M., Kusin, I. P. and Tarakanov, R. Z., 1961. Some results of the detailed study of the south Kuril Islands seismicity. *Ann. Geofis.*, 14: 119–136.

Fisher, R. L. and Hess, H. H., 1963. Trenches. In: M. N. Hill (Editor), *The Sea*. Interscience, New York, N.Y., 3: 411–436.

Fisher, R. L. and Raitt, R. W., 1962. Topography and structure of the Peru–Chile Trench. *Deep-Sea Res.*, 9: 423–443.

Fitch, T. J. and Molnar, P., 1970. Focal mechanisms along inclined earthquake zones in the Indonesia–Philippine region. *J. Geophys. Res.*, 75: 1431–1444.

Forbush, S. E., Aldrich, L. T., Tuve, M. A., Casaverde, M., Giesecke, A. A., Salgueiro, R., del Pozo, S. and Schmucker, U., 1967. Electrical conductivity anomaly in the mantle and crust under the

Andes. *Upper Mantle Project, U.S. Prog. Rept.,* 84 pp.

Foster, T. D., 1963. Heat flow measurements in the northeast Pacific and in the Bering Sea. *J. Geophys. Res.,* 67: 2990–2993.

Foster, T. D., 1969. Convection in a variable viscosity fluid heated from within. *J. Geophys. Res.,* 74: 685–693.

Francheteau, J., Harrison, C. G. A., Sclater, J. G. and Richards, M. L., 1970. Magnetization of Pacific seamounts: preliminary polar curve for the northeastern Pacific. *J. Geophys. Res.,* 75: 2035–2061.

Fromm, J. E., 1965. Numerical solutions of the non-linear equations for a heated fluid layer. *Phys. Fluids,* 8: 1757–1769.

Fujisawa, H., 1968. Temperature and discontinuities in the transition layer within the earth's mantle. Geophysical application of the olivine–spinel transition in the $Mg_2SiO_4 - Fe_2SiO_4$ system. *J. Geophys. Res.,* 73: 3281–3294.

Fujita, N., 1966. *G.S.I. Airborne Magnetometer and Geomagnetic Studies on Aeromagnetic Survey.* Thesis, Univ. of Tokyo.

Fujita, Y., 1960. On the laws concerning to the paleogeography and geotectonics of the Green Tuff geosyncline in northeastern Japan. *Earth Sci.,* 50/51: 22–35. (In Japanese with English abstract.)

Fukao, Y., 1969. On the radiative heat transfer and the thermal conductivity in the upper mantle. *Bull. Earthquake Res. Inst.,* 47: 549–569.

Fukao, Y., Mizutani, H. and Uyeda, S., 1968. Optical absorption spectra at high-temperature and radiative thermal conductivity of olivines. *Phys. Earth Planet. Inter.,* 1: 57–62.

Fukutomi, T., 1936. *Physics of Hot Springs.* Iwanami, Tokyo. (In Japanese)

Fukutomi, T., 1961. Rates of discharge of heat energy from the principal hot spring localities in Hokkaido, Japan. *J. Fac. Sci., Hokkaido Univ. (Geophys.),* 1: 315–330.

Gainanov, A. G., 1955. Pendulum gravity measurements in the Sea of Okhotsk and the northwestern Pacific. *Tr. Akad. Nauk S.S.S.R., Inst. Okeanol.,* 12: 145. (In Russian.)

Gainanov, A. G., Zverev, S. M., Kosminskaya, I. P., Livshitz, M. Kh., Milashin, A. P., Solouiv, O. N., Stroev, P. A., Sichev, P. M., Tuyezov, I. K., Tulina, Yu. V. and Fotiadi, E. E., 1968. The crust and upper mantle in the transition zone from the Pacific Ocean to the Asiatic continent. In: L. Knopoff, C. Drake and P. Hart (Editors), *The Crust and Upper Mantle of the Pacific Area –* Am. Geophys. Union, Geophys. Monogr., 12: 367–378.

Geographical Survey Institute, 1955. Gravity survey in Japan, 1. Gravity survey in Hokkaido district. *Bull. Geogr. Surv. Inst.,* 4: 23.

Geographical Survey Institute of Japan, 1957a. *Map of Japan and its Environs, 1: 2,500,000.*

Geographical Survey Institute, 1957b. Gravity survey in Japan, 2. Gravity survey in Tohoku district. *Bull. Geogr. Surv. Inst.,* 5: 1–94.

Geographical Survey Institute, 1964. Gravity survey in Japan, 3. Gravity survey in Kanto and Chubu districts. *Bull. Geogr. Surv. Inst.,* 9: 155–332.

Geographical Survey Institute, 1965. Gravity survey in Japan, 4. Gravity survey in Chubu, Kinki and Chugoku districts. *Bull. Geogr. Surv. Inst.,* 10: 55–183.

Geographical Survey Institute, 1966. Gravity survey in Japan, 5. Gravity survey in Shikoku district. *Bull. Geogr. Surv. Inst.,* 11: 59.

Geological Survey of Japan, 1968. *Geologic Map of Japan, 1: 2,000,000.* Geol. Surv. Japan, Kawasaki.

Geological Survey of Korea, 1956. *Geologic Map of Korea, 1: 1,000,000.*

Geological Survey of Manchuria, 1940. *Geologic Map of Manchuria and Adjacent Areas, 1: 3,000,000.* (Compiled by R. Saito.) Geol. Surv. Manchuria, Rept. 100.

Gilvarry, J. J., 1957. Temperature in the earth's interior. *J. Atmos. Terr. Phys.,* 10: 84–95.

Girdler, R. W., 1962. Initiation of continental drift. *Nature,* 144: 521–523.

Girdler, R. W., 1963. Geophysical studies of rift valleys. In: L. H. Ahrens, F. Press, K. Rankama and S. K. Runcorn (Editors), *Physics and Chemistry of the Earth.* Pergamon, London, 5: 122–156.

Girdler, R. W., 1964. How genuine is the circum-Pacific Belt. *Geophys. J. Roy. Astron. Soc. London,* 8: 537–540.

Gordon, R. B., 1965. Diffusion creep in the earth's mantle. *J. Geophys. Res.* 70: 2413–2418.

Goryatchev, A. V., 1962. On the relationship between geotectonic and geophysical phenomena of the Kuril–Kamtchatka folding zone at the junction zone of the Asiatic continent with the Pacific Ocean. In: G. A. MacDonald and H. Kuno (Editors), *The Crust of the Pacific Basin – Am. Geophys. Union, Geophys. Monogr.,* 6: 41–49.

Graham, J., 1949. The stability and significance of magnetism in sedimentary rocks. *J. Geophys. Res.,* 54: 131–167.

Green, D. H. and Ringwood, A. E., 1963. Mineral assemblages in a model mantle composition. *J. Geophys. Res.,* 68: 937–945.

Green, J. and Poldervaart, A., 1955. Some basaltic provinces. *Geochim. Cosmochim. Acta,* 7: ·177–188.

Griggs, D., 1939. A theory of mountain-building. *Am. J. Sci.,* 237: 611–650.

Griggs, D., 1967. Hydrolytic weakening of quartz and other silicates. *Geophys. J. Roy. Astron. Soc. London,* 14: 19–31.

Griggs, D. and Baker, D. W., 1969. The origin of deep-focus earthquakes. In: H. Mark and S. Fernbach (Editors), *Properties of Matter under Unusual Conditions.* Interscience, New York, N.Y., pp.23–42.

Griggs, D. and Handin, J., 1960. Observation on fracture and hypothesis of earthquakes. *Geol. Soc. Am., Mem.,* 79: 347–364.

Grim, P. J. and Erickson, B. H., 1969. Fracture zones and magnetic anomalies south of the Aleutian Trench. *J. Geophys. Res.,* 74: 1488–1494.

Gutenberg, B. and Richter, C. F., 1954. *Seismicity of the Earth.* Princeton Univ. Press, Princeton, N.J.

Hagiwara, T. and Rikitake, T., 1967. Japanese program on earthquake prediction. *Science,* 157: 761–768.

Hagiwara, Y., 1967. Analysis of gravity values in Japan. *Bull. Earthquake Res. Inst.,* 45: 1091–1228.

Hamilton, W., 1969. Mesozoic California and the underflow of Pacific mantle. *Geol. Soc. Am., Bull.,* ͺ80: 2409–2430.

Harada, T., 1966. Universal program to be used with an electronic computer for net-adjustment of any geodetic figure. *Bull. Geogr. Surv. Inst.,* 12: 21–39.

Harada, T., 1967. Precise readjustment of old and new first order triangulations, and the result in relation witl. destructive earthquakes in Japan. *Bull. Geogr. Surv. Inst.,* 12: 1–60.

Harada, T. and Isawa, N., 1969. Horizontal deformation of the crust in Japan. Results obtained by multiple fixed stations. *J. Geod. Soc. Japan,* 14: 101–105. (In Japanese with English abstract.)

Harata, T., 1964. The Muro Group in the Kii Peninsula, southwest Japan. *Mem. Coll. Sci., Univ. Kyoto, Ser. B,* 31: 71–94.

Harata, T., 1965. Some directional structures in the flysch-like beds of the Shimanto terrain in the Kii Peninsula, southwest Japan. *Mem. Coll. Sci., Univ. Kyoto, Ser B,* 32: 103–176.

Harris, P. G., 1957. Zone refining and the origin of potassic basalts. *Geochim. Cosmochim. Acta,* 12: 195–208.

Hasebe, K., Fujii, N. and Uyeda, S., 1970. Thermal processes under island arcs. *Tectonophysics,* 10(1–3): 335–355.

Hashimoto, I., 1966. The Mesozoic strata of uncertain ages, north of Saeki, Oita prefecture. *Rept. Earth Sci., Dept. Gen. Educ. Kyushu Univ.,* 13: 15–24. (In Japanese with English abstract.)

Hashimoto, M., Igi, S., Seki, Y., Banno, Sh. and Kojima, G., 1970. *Notes on the Metamorphic Facies Map of Japan.* Geol. Surv. Japan, Kawasaki.

Hatori, K., Kaizuka, S., Naruse, Y., Ota, Y., Sugimura, A. and Yoshikawa, T., 1964. Quaternary tectonic map of Japan (a preliminary note). *J. Geod. Soc. Japan,* 10: 111–115.

Hattori, H., 1968. Late Mesozoic to Recent tectonogenesis and its bearing on the metamorphism in New Zealand and in Japan. *Geol. Surv. Japan, Rept.,* 229: 1–45.

Hayase, I. and Ishizaka, K., 1967. Rb–Sr dating on the rocks in Japan, 1. Southwestern Japan. *J.*

Japan. Assoc. Petrol. Mineral. Econ. Geol., 58: 201–212. (In Japanese with English abstract.)

Hayes, D. E. and Heirtzler, J., 1968. Magnetic anomalies and their relation to the Aleutian island arc. *J. Geophys. Res.,* 73: 4637–4646.

Hayes, D. E. and Pitman III, W. C., 1970. Magnetic lineations in the north Pacific. *Geol. Soc. Am., Mem.,* 126: 291–314.

Heezen, B. C., 1967. Sub-oceanic trenches. In: S. K. Runcorn (Editor), *International Dictionary of Geophysics.* Pergamon, London, pp.1475–1480.

Heirtzler, J., Dickson, G. O., Herron, E. M., Pitman III, W. C. and Le Pichon, X., 1968. Marine magnetic anomalies, geomagnetics, field reversals and motions of the ocean floor and continents. *J. Geophys. Res.,* 73: 2119–2136.

Heiskanen, W. A., 1945. The gravity anomalies on the Japanese Islands and in the waters east of them. *Ann. Acad. Sci. Fenn., Ser. A III, Geol. Geogr.,* 8: 3–22.

Heiskanen, W. A. and Vening Meinesz, F. A., 1958. *The Earth and its Gravity Field.* McGraw-Hill, New York, N.Y., 470 pp.

Hermance, J. G. and Garland, G. D., 1968. Deep electrical structure under Iceland. *J. Geophys Res.,* 73: 3797–3800.

Hess, H. H., 1937. Island arcs, gravity anomalies and serpentine intrusions. *Proc. Intern. Geol. Congr., 17th, Moscow,* 2: 263–283.

Hess, H. H., 1948. Major structural features of the western north Pacific; an interpretation of H.O. 5485 Bathymetric Chart, Korea to New Guinea. *Bull. Geol. Soc. Am.,* 59: 417–446.

Hess, H. H., 1955. Serpentines, orogeny and epeirogeny. In: A. Poldervaart (Editor), *The Crust of the Earth – Geol. Soc. Am., Spec. Papers,* 62: 391–407.

Hess, H. H., 1962. History of ocean basins. In: A. E. J. Engele, H. L. James and B. F. Leonard (Editors), *Petrologic Studies – Volume in Honor of A. F. Buddington.* Geol. Soc. Am., Boulder, Colo., pp.599–620.

Hess, H. H., 1964. Seismic anisotropy of the uppermost mantle under oceans. *Nature,* 203: 629–631.

Hilde, T. W. C., Wageman, J. M. and Hammond, W. T., 1969. The structure of Tosa Terrace and Nankai Trough off southeastern Japan. *Deep-Sea Res.,* 16: 67–75.

Hisamoto, S., 1965. On anomaly of travel-time of S-waves observed in eastern Japan. *J. Seismol. Soc. Japan,* 18: 142–153; 195–203. (In Japanese with English abstract.)

Hobbs, W. H., 1925. The unstable middle section of the island arcs. *Verhandl. Geol. Mijnbouwk. Genoot.,* 8.

Hodgson, J. H., 1962. Movements in the earth's crust as indicated by earthquakes. In: S. K. Runcorn (Editor), *Continental Drift.* Academic Press, New York, N.Y.

Holmes, A., 1929. Radioactivity and earth movements. *Trans. Geol. Soc. Glasgow,* 18: 559–606.

Holmes, A., 1965. *Principles of Physical Geology.* Thomas Nelson, Edinburgh, 1288 pp.

Honda, H., 1934a. On the amplitude of the P and S waves of deep earthquakes. *Geophys. Mag., Japan Meteorol. Agency,* 8: 153–164.

Honda, H., 1934b. On the mechanism of deep earthquakes and the stress in the deep layer of the earth crust. *Geophys. Mag., Japan Meteorol. Agency,* 8: 179–185.

Honda, H. and Masatsuka, A., 1952. On the mechanisms of the earthquakes and the stress producing them in Japan and its vicinity. *Sci. Rept. Tohoku Univ., Ser. 5, Geophys.,* 4: 42–60.

Honda, H., Masatsuka, A. and Emura, K., 1957. On the mechanism of the earthquakes and the stresses producing them in Japan and its vicinity. (Second paper.) *Sci. Rept. Tohoku Univ., Ser. 5, Geophys.,* 8: 186–205.

Honda, H., Masatsuka, A. and Ichikawa, M., 1967. On the mechanism of earthquakes and the stresses producing them in Japan and its vicinity. (Third paper.) *Geophys. Mag., Japan Meteorol. Agency,* 33: 271–279.

Horai, K., 1959. Terrestrial heat flow at Hitachi, Ibaraki prefecture, Japan. *Bull. Earthquake Res. Inst.,* 37: 571–592.

Horai, K., 1963a. Terrestrial heat flow measurements in Tohoku district, Japan. *Bull. Earthquake Res. Inst.,* 41: 137–147.

Horai, K., 1963b. Terrestrial heat flow measurements in Kyushu district, Japan. *Bull. Earthquake Res.*

Inst., 41: 149–165.

Horai, K., 1963c. Terrestrial heat flow measurements in Hokkaido district, Japan. *Bull. Earthquake Res. Inst.*, 41: 167–184.

Horai, K., 1964. Terrestrial heat flow in Japan (a summary of terrestrial heat flow measurements in Japan up to December, 1962). *Bull. Earthquake Res. Inst.*, 42: 94–132.

Horai, K. and Uyeda, S., 1963. Terrestrial heat flow in Japan. *Nature*, 199: 364–365.

Horai, K. and Uyeda, S., 1969. Terrestrial heat flow in volcanic areas. In: P. J. Hart (Editor), *The Earth's Crust and Upper Mantle – Am. Geophys. Union, Geophys. Monograph.*, 13: 95–109.

Hoshino, M., 1963. Southwest Japan Trench. *J. Marine Geol.*, 1: 10–15. (In Japanese with English abstract.)

Hotta, H., 1967. The structure of sedimentary layer in the Japan Sea. *Geophys. Bull., Hokkaido Univ.*, 18: 111–131. (In Japanese with English abstract.)

Hubbert, M. K. and Rubey, W. W., 1959. Role of fluid pressure in mechanics of overthrust faulting, 1. Mechanics of fluid-filled porous solids and its application to overthrust faulting. *Bull. Geol. Soc. Am.*, 70: 115.

Hunahashi, M., 1957. Alpine orogenic movement in Hokkaido, Japan. *J. Fac. Sci., Hokkaido Univ., Ser. IV*, 9: 415–469.

Hurley, P. M., Hughes, H., Pison Jr., W. H. and Fairbairn, H. W., 1962. Radiogenic argon and strontium diffusion parameters in biotite at low temperature obtained from Alpine Fault uplift in New Zealand. *Geochim. Cosmochim. Acta*, 26: 67–80.

Huzita, K., 1962. Tectonic development of the Median Zone (Setouti) of southwest Japan since the Miocene, with special reference to the characteristic structure of the central Kinki area. *J. Geosci., Osaka City Univ.*, 6: 103–144.

Huzita, K., 1969. Tectonic development of southwest Japan in the Quaternary Period. *J. Geosci., Osaka City Univ.*, 12: 53–70.

Ichikawa, K., 1958. Bemerkungen zum tektonischen Werdegang Südwest-japans während des Paläozoikums. *J. Inst. Polytech., Osaka City Univ.*, 3: 1–13.

Ichikawa, K., 1964. Tectonic status of the Honshu major belt in southwest Japan during the Early Mesozoic. *J. Geosci., Osaka City Univ.*, 8: 71–107.

Ichikawa, K., Murakami, N., Hase, A. and Wadatsumi, K., 1968. Late Mesozoic igneous activity in the inner side of southwest Japan. *Pacific Geol.*, 1: 97–118.

Ichikawa, M., 1961. On the mechanism of the earthquakes in and near Japan during the period 1950–1957. *Geophys. Mag., Japan Meteorol. Agency*, 30: 355–403.

Ichikawa, M., 1966. Mechanism of earthquakes in and near Japan, 1950–1962. *Papers Meteorol. Geophys.*, 16: 201–229.

Ichikawa, M., 1969. Mechanism of earthquakes in and near Japan and related problems. In: P. Hart (Editor), *The Earth's Crust and Upper Mantle – Am. Geophys. Union, Geophys. Monogr.*, 13: 160–165.

Ichikawa, M., 1970. Seismic activities at the junction of Izu-Mariana and the southwestern Honshu arcs. *Geophys. Mag., Japan Meteorol. Agency*, 35: 55–69.

Ichikawa, M., 1971. Re-analyses of mechanisms of earthquakes which occurred in and near Japan, and statistical studies on the nodal plane solutions obtained, 1926–1968. *Geophys. Mag.*, 35(3): 207–273.

Ichinohe, T. and Tanaka, Y., 1964. Characteristic movements of the earth's crust related with the activity of earthquakes. *J. Geod. Soc. Japan*, 10: 154–162.

Ichinohe, T. and Tanaka, Y., 1966. On the peculiar mode of crustal movements accompanied with the activities of shallow earthquakes. *Spec. Contrib. Geophys. Inst., Kyoto Univ.*, 6: 247–254.

Iida, K., 1959. Earthquake energy and earthquake fault. *J. Earth Sci., Nagoya Univ.*, 7(2): 98–107.

Iijima, A., 1964. The Paleogene paleogeology and paleogeography of Hokkaido. *Japan J. Geol. Geogr.*, 35: 43–55.

Iijima, A. and Kagami, H., 1961. Cainozoic tectonic development of the continental slope. *J. Geol.*

Soc. Japan, 67: 561–577. (In Japanese with English abstract.)

Ikebe, N., 1956. Cenozoic geohistory of Japan. *Proc. Pacific Sci. Congr., 8th,* 2: 446–456. (Published by the National Research Council of the Philippines, Quezon.)

Ikebe, N., 1957. Sedimentary basins in the Cenozoic Era in Japan. *Sinseidaino Kenkyu,* 24/25: 1–10. (In Japanese.)

Ikebe, N. and Chizi, M., 1969. Neogene biostratigraphy and geochronology in Japan. *Shizenshi-Kenkyu,* 1(4): 25–34.

Ikebe, N. and Huzita, K., 1966. The Rokko movements, the Plio Pleistocene crustal movements in Japan. *Quaternaria,* 8: 277–287.

Ikebe, Y., Ishiwada, Y. and Kawai, K., 1965. Petroleum geology of Japan. In: *Contributions from the Government Japan to ECAFE – Petroleum Symp., 3rd, Tokyo, Japan.* Geol. Surv. Japan, Kasawaki.

Ikebe, Y., Ishiwada, Y. and Kawai, K., 1967. Petroleum geology of Japan. *Proceedings of the Third Symposium on the Development of Petroleum Resources of Asia and the far East – Mineral Resources Development Series,* 1(26): 225–234. (Published by United Nations, New York, N.Y.)

Imamura, A. and Kishinouye, F., 1928. On the horizontal shift of the dislocation accompanied by the recent destructive earthquakes in the Kwanto district and the Tango province. *Bull. Earthquake Res. Inst.,* 5: 35–41. (In Japanese with English abstract.)

Irving, E., 1964. *Paleomagnetism.* Wiley, New York, N.Y.

Isacks, B. and Molnar, P., 1969. Mantle earthquake mechanisms and sinking of the lithosphere. *Nature,* 223: 1121–1124.

Isacks, B., Oliver, J. and Sykes, L. R., 1968. Seismology and the new global tectonics. *J. Geophys. Res.,* 73: 5855–5899.

Isacks, B., Sykes, L. R. and Oliver, J., 1969. Focal mechanisms of deep and shallow earthquakes in the Tonga–Kermadec region and the tectonics of island arcs. *Geol. Soc. Am., Bull.,* 80: 1443–1470.

Isezaki, N. and Uyeda, S., 1973. Geomagnetic anomaly of the Japan Sea. (In preparation.)

Ishida, M., 1970. Seismicity and travel time anomaly in and around Japan. *Bull. Earthquake Res. Inst.,* 48: 1023–1051.

Ishikawa, T., 1968. Nature of hot springs. Paper read before the Autumn Meeting of the Volcanological Society of Japan, October, 1968.

Ishikawa, Y. and Syono, Y., 1963. Order–disorder transformation and reverse thermo-remanent magnetism in the $FeTiO_3$–Fe_2O_3 system. *J. Phys. Chem. Solids,* 24: 517–528.

Isomi, H., 1968. *Tectonic Map of Japan.* Geol. Surv. Japan, Kawasaki.

Ito, K., Endo, S. and Kawai, N., 1969. Olivine–spinel transformation in a natural forsterite. (Preprint.)

Iwabuchi, Y., 1968. Topography of trenches east of the Japanese Islands. *J. Geol. Soc. Japan,* 74(1): 37–46. (In Japanese with English abstract.)

Jacobs, J. A., Russell, R. D. and Wilson, J. T., 1959. *Physics and Geology.* McGraw-Hill, New York, N.Y.

Jaegar, J. C., 1961. *Elasticity, Fracture and Flow.* Methuen, London.

Japan Meteorological Agency, 1958; 1966. *Catalogue of Major Earthquakes which Occurred in and near Japan, 1926–1956 and 1957–1965.* Japan Meteorol. Agency, Tokyo.

Jeffreys, H., 1952. *The Earth.* Cambridge Univ. Press, London.

Kaizuka, S., 1967. Rate of folding in the Quaternary and the present. *Geogr. Rept. Tokyo Metrop. Univ.,* 2: 1–10.

Kaizuka, S., 1968. Distribution of Quaternary folds, especially rate and axis direction in Japan. *Geogr. Rept. Tokyo Metrop. Univ.,* 3: 1–9.

Kaizuka, S. and Murata, A., 1969. The amount of crustal movement during the Neogene and the Quaternary in Japan. *Geogr. Rept. Tokyo Metrop. Univ.,* 4: 1–10.

Kaizuka, S., Machida, T., Ota, Y., Sakaguchi, Y., Sugimura, A. and Yoshikawa, T., 1963. *Geomorphology of Japan, 1.* Assoc. Geol. Collabor. Japan, Tokyo, 166 pp. (In Japanese.)

Kaizuka, S., Sugimura, A., Ota, Y., Naruse, Y., Yoshikawa, T. and Hatori, K., 1966. Quaternary tectonic map of Japan (second report). *Quaternaria,* 8: 289–295.

Kaminuma, K., 1964. Crustal structure in Japan from the phase velocity of Rayleigh waves, 3. Rayleigh waves from the Mindanao shock of September 24, 1957. *Bull. Earthquake Res. Inst.,* 42: 19–38.

Kaminuma, K., 1966. The crust and upper mantle structure in Japan, 2. Crustal structure in Japan from the phase velocity of Rayleigh waves. *Bull. Earthquake Res. Inst.,* 44: 495–510.

Kaminuma, K. and Aki, K., 1963. Crustal structure in Japan from the phase velocity of Rayleigh waves, 2. Rayleigh waves from the Aleutian shock of March 9, 1957. *Bull. Earthquake Res. Inst.,* 41: 217–241.

Kanamori, H., 1963. Study on crust–mantle structure in Japan. (1) Analysis of gravity data; (2) Interpretation of the results obtained by seismic refraction studies in connection with the study of gravity and laboratory experiments; (3) Analysis of surface wave data. *Bull. Earthquake Res. Inst.,* 41: 743–759; 761–779; 801–818.

Kanamori, H., 1967a. Attenuation of P-waves in the upper and lower mantle. *Bull. Earthquake Res. Inst.,* 45: 299–312.

Kanamori, H., 1967b. Upper mantle structure from apparent velocities of P-waves recorded at Wakayama Micro-earthquake Observatory. *Bull. Earthquake Res. Inst.,* 45: 657–678.

Kanamori, H., 1968. Travel times to Japanese stations from Longshot and their geophysical implications. *Bull. Earthquake Res. Inst.,* 46: 841–859.

Kanamori, H., 1970a. Mantle beneath the Japanese Arc. *Phys. Earth Planetary Interiors,* 3: 475–483.

Kanamori, H., 1970b. The Alaska earthquake of 1964. Radiation of long-period surface waves and source mechanism. *J. Geophys. Res.,* 75: 5029–5040.

Kanamori, H. and Abe, K., 1968. Deep structure of island arcs as revealed by surface waves. *Bull. Earthquake Res. Inst.,* 46: 1001–1025.

Kanamori, H., Fujii, N. and Mizutani, H., 1968. Thermal diffusivity measurement of rock forming minerals from 300°K to 1,100°K. *J. Geophys. Res.,* 73: 595–606.

Kaneko, Sh., 1966. Transcurrent displacement along the Median Line, southwestern Japan. *New Zealand J. Geol. Geophys.,* 9: 45–59.

Kaneoka, I. and Ozima, M., 1970. On the radiometric ages of volcanic rocks from Japan. *Bull. Volcanol. Soc. Japan,* 15(1): 10–21. (In Japanese with English abstract.)

Karig, D., 1969. Extensional zones in island arc systems. *EOS,* 50: 182.

Karig, D., 1971. Structural history of the Mariana island arc system. *Geol. Soc. Am. Bull.,* 82: 323–344.

Kasahara, K. and Sugimura, A., 1964a. Horizontal secular deformation of land deduced from retriangulation data, 1. Land deformation in central Japan. *Bull. Earthquake Res. Inst.,* 42: 479–490.

Kasahara, K. and Sugimura, A., 1964b. Spatial distribution of horizontal secular strain in Japan. *J. Geod. Soc. Japan,* 10: 139–145.

Kaseno, Y., Sakamoto, T. and Ishida, S., 1961. A contribution to the Neogene history of the eastern Hokuriku district, central Japan. In: S. Matsushita (Editor), *Professor Jiro Makiyama Memorial Volume.* Univ. Kyoto, Kyoto, pp.83–95. (In Japanese with English abstract.)

Kato, Y., 1968. Northeastern Japan anomaly of the upper mantle. *Conductivity Anomaly Symp., Earthquake Res. Inst.,* pp.19–31.

Katsumata, M., 1960. The effect of seismic zones upon transmission of seismic waves. *Quart. J. Seismol.,* 25: 89–95.

Katsumata, M. and Sykes, L. R., 1969. Seismicity and tectonics of the western Pacific: Izu–Mariana– Caroline and Ryukyu–Taiwan regions. *J. Geophys. Res.,* 74: 5923–5948.

Kawachi, Sh. and Kawachi, Y., 1963. Volcanisms of the Kirigamine and the Ara fune areas after the Pliocene. *Earth Sci.,* 64: 1–7; and 65; 33–37. (In Japanese with English abstract.)

Kawada, K., 1964. Variation of the thermal conductivity of rocks, 1. *Bull. Earthquake Res. Inst.,* 42: 631–647.

Kawada, K., 1967. Variation of the thermal conductivity of rocks, 2. *Bull. Earthquake Res. Inst.*, 44: 1071–1091.

Kawai, K., 1965. Natural gas geology of the southern Kanto region. Japan. In: *Contributions from the Government of Japan to ECAFE, Agenda Item VI*, 1–18. Geol. Surv. Japan, Kawasaki.

Kawai, K., 1967. Natural gas geology of the southern Kanto region, Japan. Proceedings of the Third Symposium on the Development of Petroleum Resources of Asia and the Far East Mineral Resources Development Series, 1(26): 235–249. (Published by United Nations, New York, N.Y.)

Kawai, N. and Nakajima, T., 1970. The evolution of the island arc of Japan. In: S. Uyeda (Editor), *Thermal Structures under Japanese Islands*. Research Group of Thermal Structures under Japanese Islands, Tokyo, pp.229–275.

Kawai, N., Ito, H. and Kume, S., 1961. Deformation of the Japanese Islands as inferred from rock magnetism. *Geophys. J. Roy. Astron. Soc. London*, 6: 124–130.

Kawai, N., Hirooka, H. and Nakajima, T., 1969. Palaeomagnetic and potassium–argon age informations supporting Cretaceous–Tertiary hypothetic bend of the main island Japan. *Palaeogeogr., Palaeoclimatol., Palaeoecol.*, 6: 277–282.

Kawano, Y. and Ueda, Y., 1966. K–A dating on the igneous rocks in Japan, 5. *J. Japan. Assoc. Petrol. Mineral. Econ. Geol.*, 56: 191–211. (In Japanese with English abstract.)

Kawano, Y. and Ueda, Y., 1967. Periods of the igneous activities of the granitic rocks in Japan by K–A dating method. *Tectonophysics*, 4: 523–530.

Kay, M., 1951. North American geosynclines. *Geol. Soc. Am., Mem.*, 48: 1–148.

Kennedy, W. Q., 1933. Trends of differentiation in basaltic magmas. *Am. J. Sci.*, 25: 239–256.

Kimura, T., 1966. Thickness distribution of sandstone beds and cyclic sedimentation in the turbidite sequences at two localities in Japan. *Bull. Earthquake Res. Inst.*, 44: 561–607.

Kimura, T., 1967. Geologic structure of the Shimanto Group in the southern Oigawa district, central Japan, analyzed by the minor geologic structures. *Commem. Publ. Prof. Y Sasa*, pp. 21–48. (In Japanese with English abstract.)

Kimura, T., 1968. Some folded structures and their distribution in Japan. *Japan J. Geol. Geogr.*, 39: 1–26.

Kishu Shimanto Research Group, 1968. The study of the Shimanto terrain in the Kii Peninsula, southwest Japan, 2. *Earth Sci.*, 22: 224–231. (In Japanese with English abstract.)

Kitamura, N., 1963. Tertiary tectonic movements in the Green Tuff area. *Kaseki (Fossil)*, 5: 123–137. (In Japanese.)

Knopoff, I., 1964. The convection current hypothesis. *Rev. Geophys.*, 2: 89–122.

Kobayashi, T., 1941. The Sakawa orogenic cycle and its bearing on the origin of the Japanese Islands. *J. Fac. Sci., Imp. Univ. Tokyo, Sect.2*, 5: 219–578.

Kobayashi, T., 1955. Up and down movements now in action in Japan. *Geol. Rundschau*, 43: 233–247.

Kobayashi, T., 1956. The insular arc of Japan; its hinter basin and its linking with the peri-Tunghai arc. *Proc. Pacific Sci. Congr., 8th*, II-A: 799–808.

Konishi, K., 1965. Geotectonic framework of the Ryukyu Islands (Nansei-shoto). *J. Geol. Soc. Japan*, 71: 437–457. (In Japanese with English abstract.)

Kosminskaya, I. P. and Zverev, S. M., 1968. Deep seismic sounding in the transition zones. In: L. Knopoff, C. Drake and P. Hart (Editors), *The Crust and Upper Mantle of the Pacific Area* – Am. Geophys. Union, *Geophys. Monogr.*, 12: 122–130.

Kosminskaya, I. P., Zverev, S. M., Veitsman, P. S., Tulina, Yu.V. and Krakshina, R. M., 1963. Basic feature of the Sea of Okhotsk and the Kuril–Kamchatka zone of the Pacific Ocean from deep seismic sounding data. *Izv. Akad. Nauk S.S.S.R., Geophys. Ser.*, 1: 20–41.

Kovylin, V. N., 1966. Results of seismic research in the southwestern part of the abyssal basin of the Sea of Japan. *Oceanology*, 6: 294–305.

Kovylin, V. N. and Neprochnov, Yu.P., 1965. Crustal structure and sedimentary layer in the Sea of Japan derived from seismic data. *Izv. Akad. Nauk S.S.S.R., Geol. Ser.*, 4: 10–26.

Kuenen, Ph.H., 1936. The negative isostatic anomalies in the East Indies (with experiments). *Leidsche*

Geol. Mededel., 8: 169—214.

Kumazawa, M., 1963. A fundamental thermodynamic theory non-hydrostatic field and the stability of mineral orientation and phase equilibrium. *J. Earth Sci., Nagoya Univ.*, 11: 145—217.

Kumazawa, M., 1967. The elastic anisotropy of the upper mantle and its geotectonic significance. (Unpublished.)

Kumazawa, M. and Tada, T., 1964. Anisotropy of the upper mantle in velocity of P-waves and its significance. Autumn Meeting Seismol. Soc. Japan, 1964. (Unpublished.)

Kuno, H., 1952. Cenozoic volcanic activity in Japan and surrounding areas. *Trans. N.Y. Acad. Sci., Ser. II*, 14: 225—231.

Kuno, H., 1959. Origin of Cenozoic petrographic provinces of Japan and surrounding areas. *Bull. Volcanol.*, 20: 37—76.

Kuno, H., 1960. High-alumina basalt. *J. Petrol.*, 1: 121—145.

Kuno, H., 1962. *Catalogue of the Active Volcanoes of the World Including Solfatara Fields, 11. Japan, Taiwan and Marianas*. Intern. Volcanol. Assoc., Napoli, 332 pp.

Kuno, H., 1964. Differences in chemical composition and mechanism of generation between the basalt magmas in Japan and Hawaii. *J. Geograph. Tokyo*, 73: 279—280. (In Japanese with English abstract.)

Kuno, H., 1966. Lateral variation of basalt magma type across continental margins and island arcs. *Bull. Volcanol.*, 29: 195—222.

Kuno, H., 1968. Differentiation of basalt magmas. In: H. H. Hess and A. Poldervaart (Editors), *Basalts: the Poldervaart Treatise on Rocks of Basaltic Composition*. Interscience, New. York, N.Y., 2: 623—688.

Kuno, H., Yamasaki, K., Iida, H. and Nagashima, K., 1957. Differentiation of Hawaiian magmas. *Japan J. Geol. Geogr.*, 28: 179—218.

Kushiro, I., 1965. The liquidus relations in the systems forsterite—$CaAl_2SiO_6$—silica and forsterite—nepheline—silica at high pressures. *Carnegie Inst., Ann. Rept. Dir. Geophys. Lab.*, 1964/1965: 103—109.

Kushiro, I., 1966. Problem of H_2O in the mantle. *Bull. Volcanol. Soc. Japan*, 11: 116—126. (In Japanese with English abstract.)

Kushiro, I., 1968. Compositions of magmas formed by partial zone melting of the earth's upper mantle. *J. Geophys. Res.*, 73: 619—634.

Kushiro, I., 1969. The system forsterite—diopside—silica with and without water at high pressures. *Am. J. Sci. (Shairer Vol.)*, 267-A: 269—294.

Kushiro, I. and Kuno, H., 1963. Origin of primary basalt magmas and classification of basaltic rocks. *J. Petrol.*, 4: 75—89.

Kushiro, I., Syono, Y. and Akimoto, S., 1967. Stability of phlogopite at high pressures and possible presence of phlogopite in the earth's upper mantle. *Earth Planetary Sci. Lett.*, 3: 197—203.

Kushiro, I., Syono, Y. and Akimoto, S., 1968a. Melting of a peridotite nodule at high pressures and high water pressures. *J. Geophys. Res.*, 73: 6023—6029.

Kushiro, I., Yoder, H. S. and Nishikawa, M., 1968b. Effect of water on melting of enstatite. *Geol. Soc. Am., Bull.*, 79: 1685—1692.

Kuwahara, T., 1968. The Noobi Basin and its fault block movements. *The Quaternary Res.*, 7: 235—247. (In Japanese with English abstract.)

Lake, Ph., 1931. Island arcs and mountain building. *Geogr. J.*, 78: 149—160.

Langseth, M. G., Le Pichon, X. and Ewing, M., 1966. Crustal structure of the mid-oceanic ridges, 5. Heat flow through the Atlantic Ocean floor and convection currents. *J. Geophys. Res.*, 71: 5321—5355.

Lawson, A. C., 1932. Insular arcs, foredeeps and geosynclinal seas of the Asiatic coast. *Bull. Geol. Soc. Am.*, 43: 353—382.

Lee, W. H. K. and Uyeda, S., 1965. Review of heat flow data. In: W. H. K. Lee (Editor), *Terrestrial Heat Flow — Am. Geophys. Union, Geophys. Monogr.*, 8: 87—190.

Le Pichon, X., 1968. Sea-floor spreading and continental drift. *J. Geophys. Res.*, 73: 3661—3697.

Lubimova, E. A., 1958. Thermal history of the earth with consideration of variable thermal conductivity of the mantle. *Geophys. J. Roy. Astron. Soc. London*, 1: 115–135.

Lubimova, E. A., 1963. Heat transfer by excitons on the earth's mantle. In: V. A. Magnitzky (Editor), *Problems of Theoretical Seismology and Physics of the Earth's Interior*. (English translation published by National Science Foundation and National Aeronautics and Space Administration, pp.60–65.)

Lubimova, E. A. and Magnitzky, V. A., 1964. Thermo-elastic stresses and the energy of earthquakes. *J. Geophys. Res.*, 69: 3443–3447.

Ludwig, W. J., Ewing, J. I., Ewing, M., Murauchi, S., Den, N., Asano, S., Hotta, H., Hayakawa, M., Ichikawa, K. and Noguchi, I., 1966. Sediments and structure of the Japan Trench. *J. Geophys. Res.*, 71: 2121–2137.

Lyustikh, E. N., 1967. Criticism of hypotheses of convection and continental drift. *Geophys. J.*, 14: 347–352.

MacDonald, G. J. F., 1963. Deep structure of continents. *Rev. Geophys.*, 1: 587–665.

MacDonald, G. J. F., 1965. Geophysical deductions from observations of heat flow. In: W. H. K. Lee (Editor), *Terrestrial Heat Flow – Am. Geophys. Union, Geophys. Monogr.*, 8: 191–210.

Machida, H. and Suzuki, M., 1971. Absolute ages of volcanic ash and Late Quaternary chronology. *Kagaku (Natural Science)*, 41(5): 263–270. (In Japanese.)

Magnitzky, V. A., 1964. Zone melting as a process of the crust formation. *Izv. Akad. Nauk S.S.S.R., Geol. Ser.*, 11: 3–8. (In Russian.)

Maritime Safety Board, Japan, 1966. *Bathymetric Chart of the Adjacent Seas of Nippon, 1: 3,000,000.*

Mason, D., 1958. Magnetic survey off the west coast of the United States, between the latitudes 32°N and 36°N and longitudes 121°W and 128°W. Geophys. *J. Roy. Astron. Soc. London*, 1: 320–329.

Mason, R. G. and Raff, A. D., 1961. Magnetic survey off the west coast of North America, 32°N–42°N latitude. *Bull. Geol. Soc. Am.*, 72: 1259–1266.

Matsuda, T., 1962. Crustal deformation and igneous activity in the South Fossa Magna, Japan. In: L. Knopoff, C. Drake and H. Kuno (Editors), *The Crust of the Pacific Basin – Am. Geophys. Union, Geophys. Monogr.*, 6: 140–150.

Matsuda, T., 1964. Island arc features and the Japanese Islands. *J. Geograph. Tokyo*, 73: 271–278. (In Japanese with English abstract.)

Matsuda, T., 1967. Seismogeology. *Zisin, J. Seismol. Soc. Japan*, 20: 230–235. (In Japanese.)

Matsuda, T., 1969. Active faults and great earthquakes. *Kagaku (Nat. Sci.)*, 39: 398–407. (In Japanese.)

Matsuda, T. and Kuriyagawa, S., 1965. Lower grade metamorphism in the eastern Akaishi Mountains, central Japan. *Bull. Earthquake Res. Inst.*, 43: 209–235. (in Japanese with English abstract.)

Matsuda, T. and Uyeda, S., 1971. On the Pacific-type orogeny and its model. Extension of the paired belts concept and possible origin of marginal seas. *Tectonophysics*, 11: 5–27.

Matsuda, T., Nakamura, K. and Sugimura, A., 1966. Late Cenozoic orogeny in Japan. *Pacific Sci. Congr., 11th, Tokyo.*

Matsuda, T., Nakamura, K. and Sugimura, A., 1967. Late Cenozoic orogeny in Japan. *Tectonophysics*, 4: 349–366.

Matsumoto, Takashi., 1968. A hypothesis on the origin of the Late Mesozoic volcano-plutonic association in east Asia. *Pacific Geol.*, 1: 77–83.

Matsumoto, Takashi. and Wadatsumi, K., 1959. Cenozoic structural development in northern Tazima. *J. Geol. Soc. Japan*, 65: 117–127. (In Japanese with English abstract.)

Matsumoto, Tatsuro., 1967. Fundamental problems in the circum-Pacific orogenesis. *Tectonophysics*, 4(4–6): 595–613.

Matsuzaki, T., 1966. Magnetic anomalies over and around the Japan Trench off Sanriku and the Yamato Bank in the Sea of Japan. *Rept. Hydrogr. Res.*, pp. 1–10.

Matuyama, M., 1929. On the direction of magnetization of basalt in Japan, Tyosen and Manchuria.

Proc. Imp. Acad., Tokyo, 5: 203–205.

Matuyama, M., 1936. Distribution of gravity over the Nippon Trench and related areas. *Proc. Imp. Acad. Tokyo,* 12: 93–95.

Matuzawa, T., 1953. Feldtheorie der Erdbeben. *Bull. Earthquake Res. Inst.,* 31: 179–201.

Matuzawa, T., Yamada, K. and Suzuki, T., 1929. On the forerunners of earthquake motions (second paper). *Bull. Earthquake Res. Inst.,* 7: 241–260.

Maxwell, A. E., von Herzen, R. P., Hsu, K. J., Andrews, J. E., Saito, T., Percival Jr., S. F., Milow, E. D. and Boyce, R. E., 1970. Deep sea drilling in the South Atlantic. *Science,* 168: 1047–1059.

MacBirney, A. R. (Editor), 1969. *Proceedings of the Andesite Conference, Eugene and Bond. Ore., U.S.A.*

McBirney, A. R. and Gass, I. G., 1967. Relations of oceanic volcanic rocks to mid-oceanic rises and heat flow. *Earth Planetary Sci. Lett.,* 2: 265–276.

McConnell Jr., R. K., 1965. Isostatic adjustment in a layered earth. *J. Geophys. Res.,* 70: 5171–5188.

McConnell Jr., R. K., McClaine, L. A., Lee, D. W., Aronson, J. R. and Allen, R. V., 1967. A model for planetary igneous differentiation. *Rev. Geophys.,* 5: 121–172.

McKenzie, D. P., 1967a. Some remarks on heat flow and gravity anomalies. *J. Geophys. Res.,* 72: 6261–6273.

McKenzie, D. P., 1967b. The viscosity of the mantle. *Geophys. J. Roy. Astron. Soc. London,* 14: 297–305.

McKenzie, D. P., 1968. The geophysical importance of high-temperature creep. In: R. A. Phinney (Editor), *The History of the Earth's Crust.* Princeton Univ. Press, Princeton, N.J.

McKenzie, D. P., 1969. Speculations on the consequences and causes of plate motions. *Geophys. J. Roy. Astron. Soc. London,* 18: 1–37.

McKenzie, D. P. and Morgan, W. J., 1969. Evolution of triple junctions. *Nature,* 224: 125–133.

McKenzie, D. P. and Parker, R. L., 1967. The north Pacific: an example of tectonics on a sphere. *Nature,* 216: 1276–1280.

McKenzie, D. P. and Sclater, J. G., 1968. Heat flow inside the island arcs of the northwestern Pacific. *J. Geophys. Res.,* 73: 3173–3179.

Menard, H. W., 1964. *Marine Geology of the Pacific.* McGraw-Hill, New York, N.Y., 271 pp.

Menard, H. W., 1965. Sea-floor relief and mantle convection. In: L. H. Ahrens, F. Press, K. Rankama and S. K. Runcorn (Editors), *Physics and Chemistry of the Earth.* Pergamon, London, 6: 315–364.

Mikumo, T., 1966. A study on crustal structure in Japan by the use of seismic and gravity data. *Bull. Earthquake Res. Inst.,* 44: 965–1007.

Miller, J., Shido, F., Banno, S. and Uyeda, S., 1961. New data on the age of orogeny and metamorphism in Japan. *Japan J. Geol. Geogr.,* 32: 145–151.

Minato, M., 1952. Problem of the Green Tuff area. *Sinseidaino Kenkyu (Study of the Cenozoic Era),* 14: 239–248. (In Japanese.)

Minato, M., Yagi, K. and Hunahashi, M., 1956. Geotectonic synthesis of the Green Tuff region in Japan. *Bull. Earthquake Res. Inst.,* 34: 237–264.

Minato, M., Gorai, M. and Hunahashi, M. (Editors), 1965. *The Geologic Development of the Japanese Islands.* Tsukiji Shokan, Tokyo, 442 pp.

Minear, J. W. and Toksöz, M. N., 1969. Thermal regime of a downgoing slab and new global tectonics. *J. Geophys. Res.,* 75: 1397–1419.

Ministry of Geology of the U.S.S.R., 1965. *Geologic Map of the U.S.S.R., Scale 1:2,500,000.* (In Russian.)

Mitchell, A. H. G., 1966. Geology of South Malekula. *New Hebrides Condominium Geol. Surv., Rept.* Port Vila, New Hebrides, 42 pp.

Miyabe, N., 1952. Vertical earth movements in Japan as deduced from the results of rerunning the precise levels. *Bull. Earthquake Res. Inst.,* 30: 127–162.

Miyabe, N., Miyamura, S. and Mizoue, M., 1966. A map of secular vertical movements of the earth's crust in Japan. *Ann. Acad. Sci. Fenn., Ser. A. III. Geol. Geogr.,* pp.287–289.

Miyamura, S., 1962. Types of crustal movements accompanied with earthquakes. In: O. Meissner (Editor), *I.U.G.G., I.A.G., C.R.C.M. Intern. Symp., 1st, Leipzig, 1962 – Recent Crustal Move-*

ments. Akademie-Verlag, Berlin, pp.235–251.

Miyamura, S. and Mizoue, M., 1964. Secular vertical movements of the earth's crust in Japan (modes of movements in space and time with special reference to geotectonics). *J. Geod. Soc. Japan*, 10: 123–138.

Miyamura, S., Hori, M. and Matumoto, H., 1966. Local earthquakes in Kii Peninsula, 5. Development of micro-earthquake observation network in Kii Peninsula during 1957–1962 and observed seismicity. *Bull. Earthquake Res. Inst.*, 44: 709–729. (In Japanese with English abstract.)

Miyashiro, A., 1959. Abukuma, Ryoke and Sanbagawa metamorphic belts. *J. Geol. Soc. Japan*, 65(769): 624–637. (In Japanese with English abstract.)

Miyashiro, A., 1961a. Evolution of metamorphic belts. *J. Petrol.*, 2: 277–311.

Miyashiro, A., 1961b. Rock metamorphism. In: C. Tsuboi (Editor), *Constitution of the Earth*. Iwanami Shoten, Tokyo. (In Japanese.)

Miyashiro, A., 1966. Some aspects of peridotite and serpentinite in orogenic belts. *Japan J. Geol. Geogr.*, 37: 45–61.

Miyashiro, A., 1967. Orogeny, regional metamorphism and magmatism in the Japanese Islands. *Medd. Dansk Geol. Foren.*, 17: 390–446.

Mizoue, M., 1968. Types of crustal movements found by precise releveling. *Mem. Geol. Soc. Japan*, 2: 9–14. (In Japanese with English abstract.)

Mizutani, H., Baba, K., Kobayashi, N., Chang, C. C., Lee, C. H. and Kang, Y. S., 1970. Heat flow in Korea. *Tectonophysics*, 10: 183–203.

Mochizuki, K., 1943. *Theory of Tectonic Features in East Asia*. Kokon Shoin, Tokyo, 200 pp. (In Japanese.)

Mogi, A., 1969. Topographie sous-marine de mers bordières du Pacifique d'ouest. *La Mer*, 7: 213–219. (In Japanese with French résumé.)

Mogi, K., 1958. Relation between the eruptions of various volcanoes and the deformations of the ground surface around them. *Bull. Earthquake Res. Inst.*, 36: 99–134.

Mogi, K., 1967. Regional variation of aftershock activity. *Bull. Earthquake Res. Inst.*, 45: 711–726.

Mogi, K., 1970. Recent horizontal deformation of the earth's crust and tectonic activity in Japan, 1. *Bull. Earthquake Res. Inst.*, 48(3): 413–430.

Molengraaff, G. A. F., 1914. Folded mountain chains, overthrust sheets and block-faulted mountains in the East Indian Archipelago. *Intern. Geol. Congr., 12th, Ottawa, Compt. Rend.*, pp.689–702.

Molnar, P. and Sykes, L. R., 1969. Tectonics of the Caribbean and Middle America regions from focal mechanisms and seismicity. *Geol. Soc. Am., Bull.*, 80: 1639–1684.

Momose, K., 1963. Studies on the variations of the earth's magnetic field during Pliocene times. *Bull. Earthquake Res. Inst.*, 41: 487–534.

Morgan, W. J., 1968. Rises, trenches, great faults and crustal blocks. *J. Geophys. Res.*, 73: 1959–1982.

Murai, I., 1966. Geotectonic maps of the Japanese Islands, 1. Map of faults. *Bull. Earthquake Res. Inst.*, 44: 551–559.

Muramatu, I., Kazita, S., Suzuki, S. and Sugimura, A., 1964. Structure of the Midori fault, a part of the Neo Valley fault in central Japan. *Sci. Rept., Fac. Lib. Arts Educ., Gifu Univ. (Nat. Sci.)*, 3: 308–317. (In Japanese with English abstract.)

Murauchi, S., 1966. On the origin of the Sea of Japan. Paper read at the monthly meeting of the Earthquake Research Institute, University of Tokyo.

Murauchi, S. and Yasui, M., 1968. Geophysical investigations in the seas around Japan. *Kagaku*, 38(4). (In Japanese.)

Murauchi, S., Den, N., Asano, S., Hotta, H., Chujo, J., Asanuma, T., Ichikawa, K. and Noguchi, I., 1964. A seismic refraction exploration of Kumano Nada (Kumano Sea), Japan. *Proc. Japan Acad.*, 40: 111–115.

Murauchi, S., Den, N., Asano, S., Hotta, H., Yoshii, T., Asanuma, T., Hagiwara, K., Ichikawa, K., Sato, T., Ludwig, W. J., Ewing, J. I., Edgar, T. and Houtz, R. E., 1968. Crustal structure of the Philippine sea. *J. Geophys. Res.*, 73: 3143–3171.

Murray, H. W., 1945. Profiles of the Aleutian Trench. *Bull. Geol. Soc. Am.*, 56: 757–782.

Nafe, J. E. and Drake, C. L., 1957. Variation with depth in shallow and deep water marine sediments of porosity, density and the velocities of compressional and shear waves. *Geophysics*, 22: 523–552.

Nagao, T., 1933. "Nappes" and "klippes" in central Hokkaido. *Proc. Imp. Acad., Tokyo*, 9: 101–104.

Nagasaka, K., Francheteau, J. and Kishii, T., 1970. Terrestrial heat flow in the Celebes and Sulu seas. *Marine Geophys. Res.*, 1: 99–103.

Nagata, T., 1961. *Rock Magnetism*. Maruzen, Tokyo.

Nagata, T., 1967. Introduction to conductivity anomaly. *Symp. Conductivity Anomaly, Kakioka*, pp.1–8.

Nagata, T., Akimoto, S. and Uyeda, S., 1951. Reverse thermo-remanent magnetism. *Proc. Japan Acad.*, 27: 643–645.

Nagata, T., Akimoto, S., Shimizu, Y., Kobayashi, K. and Kuno, H., 1959. Paleomagnetic studies on Tertiary and Cretaceous rocks in Japan. *Proc. Japan Acad.*, 35: 378–383.

Nagata, T., Akimoto, S., Uyeda, S., Shimizu, Y., Ozima, M., Kobayashi, K. and Kuno, H., 1957. Paleomagnetic studies on a Quaternary volcanic region in Japan. *J. Geomagn. Geoelectr.*, 9: 23–41.

Nakagawa, H., Niitsuma, N. and Hayasaka, I., 1969. Late Cenozoic geomagnetic chronology of the Boso Peninsula. *J. Geol. Soc. Japan*, 75: 267–280. (In Japanese with English abstract.)

Nakamura, K., 1969. A hypothesis on island-arc tectonics. *Ann. Meeting, Geol. Soc. Japan, 76th, Oct. 1969, Niigata*. (Unpublished.)

Nakamura, K. and Tsuneishi, Y., 1967. Ground cracks at Matsushiro probably of underlying strike-slip fault origin, 2. The Matsushiro earthquake fault. *Bull. Earthquake Res. Inst.*, 45: 417–471.

Nakamura, K., Kasahara, K. and Matsuda, T., 1964. Tilting and uplift of an island (Awashima), near the epicentre of the Niigata earthquake in 1964. *J. Geod. Soc. Japan*, 10: 172–179.

Naruse, Y., 1959. On the formation of the paleo-Tokyo Bay – geologic history of the Late Cenozoic deposits in the southern Kanto region. 1. *The Quaternary Res.*, 1: 143–155. (In Japanese with English abstract.)

Naruse, Y., 1966. *The Quaternary Geology of Tokyo (Guidebook for Geologic Trip No. 1 at the 11th Pacific Science Congress)*. Japan Assoc. Quaternary Res., Tokyo, 25 pp.

Naruse, Y., 1968. Quaternary crustal movements in Kanto district. *Mem. Geol. Soc. Japan*, 2: 29–32. (In Japanese with English abstract.)

Naruse, Y. and Sugimura, A., 1965. Paleogeography and neotectonics. In: M. Minato, M. Gorai and M. Hunahashi (Editors), *The Geologic Development of the Japanese Islands*. Tsukiji Shokan, Tokyo, pp.358–363.

National Research Center for Disaster Prevention, 1969. *Quaternary Tectonic Map of Japan*. Natl. Res. Center for Disaster Prevention, Tokyo, 6 sheets.

Naumann, E., 1885. *Ueber den Bau und Entstehung der Japanischen Inseln*. Friedländer, Berlin, 91 pp.

Néel, L., 1949. Theorie du trainage magnétique des ferromagnétiques au grains fins avec applications aux terres cuites. *Ann. Géophys.*, 5: 99–136.

Nitani, H. and Imayoshi, B., 1963. On the analysis of the deep sea observations in the Kurile–Kamchatka Trench. *J. Oceanogr. Soc. Japan*, 19(2): 1–7.

Nomura, S., 1967. On the properties of paleomagnetic polarity of the Neogene and Quaternary volcanic rocks in Japan. *Gunma J. Lib. Arts Sci.*, 1: 11–35.

Noponen, I., 1969. Wave velocities in crust and upper 300 kilometres of mantle in western and central Japan. *Bull. Intern. Inst. Seismol. Earthquake Eng.*, 6: 11–37.

Nozawa, T., 1968. Radiometric ages of granitic rocks in the outer zone of southwest Japan and its extension; 1968 summary and north-shift hypothesis of igneous activity. *J. Geol. Soc. Japan*, 74: 485–489. (In Japanese with English abstract.)

Nozawa, T., 1970. Isotopic ages of Late Cretaceous acid rocks in the Japanese Islands: summary and notes in 1970. *J. Geol. Soc. Japan*, 76(10): 493−518. (In Japanese with English abstract.)

Officer, C. B., Ewing, J. I., Hennion, J. F., Harkrider, D. G. and Miller, D. E., 1959. Geophysical investigations in the eastern Caribbean; summary of 1955 and 1956 cruises. In: L. H. Ahrens, F. Press, K. Rankama and S. K. Runcorn (Editors), *Physics and Chemistry of the Earth*. Pergamon, London, 3: 17−109.

Okada, A., 1968. Strike-slip faulting of Late Quaternary along the Median Dislocation Line in the surroundings of Awa−Ikeda, northeastern Shikoku. *Quaternary Res.*, 7: 15−26. (In Japanese with English abstract.)

Okada, A., 1970. Fault topography and rate of faulting along the Median Tectonic Line in the drainage basin of the River Yoshino, northeastern Shikoku, Japan. *Geogr. Rev. Japan*, 43: 1−21. (In Japanese with English abstract.)

Okai, K., 1959. A perturbation method for studying thermal convection problems. *J. Phys. Earth*, 7: 1−20.

Oliver, J. and Isacks, B., 1967. Deep earthquake zones, anomalous structures in the upper mantle, and the lithosphere. *J. Geophys. Res.*, 72: 4259−4275.

Ono, Ch. and Isomi, H., 1967. Comparison of areas covered by different rocks in the Japanese Islands. *Bull. Geol. Surv. Japan*, 18: 467−476. (In Japanese with English abstract.)

Orowan, E., 1960. Mechanism of seismic faulting. *Geol. Soc. Am., Mem.*, 79: 323−345.

Orowan, E., 1965. Convection in a non-Newtonian mantle, continental drift, and mountain building. *Phil. Trans. Roy. Soc.*, 258: 284−313.

Orowan, E., 1967. Island arcs and convection. *Geophys. J. Roy. Astron. Soc. London*, 14: 385−393.

Ota, Y., 1968. Deformed shorelines and Late Quaternary crustal movements in Japan. *Mem. Geol. Soc. Japan*, 2: 15−24. (In Japanese with English abstract.)

Ota, Y., 1969. Crustal movements in the Late Quaternary considered from the deformed terrace plains in northeastern Japan. *Japan J. Geol. Geogr.*, 40: 41−61.

Otatume, K., 1941. On the overthrust sheets in the southern part of the Isikari coal-field, Hokkaido. In: R. Aoki (Editor), *Jubilee Publ. Commem. Prof. H. Yabe, M.I.A. 60th Birthday*, 2: 973−988. (In Japanese with English abstract.)

Otuka, Y., 1931. Early Pliocene crustal movement in the outer zone of southwest Japan and in the Naumann's Fossa Magna. *Bull. Earthquake Res. Inst.*, 9: 340−352.

Otuka, Y., 1932. Post-Pliocene crustal movements in the outer zone of southwest Japan and in the "Fossa Magna", 1. *Bull. Earthquake Res. Inst.*, 10: 701−722.

Otuka, Y., 1933. Contraction of the Japanese Islands since the Middle Neogene (advance paper). *Bull. Earthquake Res. Inst.*, 11: 724−731.

Otuka, Y., 1936. The earthquake of central Taiwan, April 21, 1935, and earthquake faults. *Bull. Earthquake Res. Inst.*, 13: 22−74.

Otuka, Y., 1937. Some geologic considerations of the folded Tertiary zones in Japan. *Bull. Earthquake Res. Inst.*, 15: 1041−1046.

Otuka, Y., 1938. A geologic interpretation on the underground structure of the Sitito−Mariana island arc in the Pacific Ocean. *Bull. Earthquake Res. Inst.*, 16: 201−211.

Otuka, Y., 1939. Tertiary crustal deformations in Japan with short remarks on Tertiary paleo-geography. In: R. Aoki (Editor), *Jubilee Publ. Commem. Prof. H. Yabe, M.I.A., 60th Birthday*, 1: 481−519.

Otuka, Y., 1941. Active rock folding in Japan. *Proc. Imp. Acad. Tokyo*, 17: 518−522.

Oxburgh, E. R. and Turcotte, D. L., 1968. Mid-ocean ridges and geotherm distribution during mantle convection. *J. Geophys. Res.*, 73: 2643−2662.

Ozawa, A., 1963. Neogene orogenesis, igneous activity and mineralization in the central part of north-east Japan. 1. On the Neogene igneous activity. *J. Japan Assoc. Miner. Petrol. Econ. Geol.*, 50: 167−184 (in Japanese with English abstract).

Ozima, M., Ueno, N., Shimizu, N. and Kuno, H., 1967. Rb−Sr and K−A isotopic investigations of Sidara granodiorites and the associated Ryoke metamorphic belt, central Japan. *Japan. J. Geol. Geogr.*, 38 (2−4): 159−162.

Parkinson, W. D., 1959. Directions of rapid geomagnetic fluctuations. Geophys. *J. Roy. Astron. Soc.* *London,* 2: 1–14.

Parkinson, W. D., 1962. The influence of continents and oceans on geomagnetic variations. *Geophys. J.,* 4: 441–449.

Parkinson, W. D., 1964. Conductivity anomalies in Australia and the ocean effect. *J. Geomagn. Geoelectr.,* 15: 222–226.

Pearson, J. R. A., 1958. On convection cells induced by surface tension. *J. Fluid Mech.,* 4: 489–500.

Pekeris, C. L., 1935. Thermal convection in the interior of the earth. *Monthly Not. Roy. Astron. Soc., Geophys. Suppl.,* 3: 343.

Peter, G., 1966. Magnetic anomalies and fracture pattern in the northeast Pacific Ocean. *J. Geophys. Res.,* 71: 5365–5374.

Pitman Jr., W. C. and Hayes, D. E., 1968. Sea-floor spreading in the Gulf of Alaska. *J. Geophys. Res.,* 73: 6571–6580.

Plafker, G., 1965. Tectonic deformation associated with the 1964 Alaska earthquake. *Science,* 148: 1675–1687.

Pollack, H. N., 1967. Tectonic implications of a thermal contrast between continents and oceans. *J. Geophys. Res.,* 72: 5043–5049.

Polyak, B. G., 1966. *Geothermal Characteristics of the Region of Contemporary Volcanism.* Nauka, Moscow, 179 pp. (In Russian.)

Press, F., 1960. Crustal structure in the California–Nevada region. *J. Geophys. Res.,* 65: 1039–1051.

Quaternary Research Group of the Kiso Valley and Kigoshi, K., 1964. Radiocarbon date of the Kisogawa volcanic mud flows and its significance on the Würmian chronology of Japan. *Earth Sci.,* 71: 1–7.

Raff, A. D. and Mason, R. G., 1961. Magnetic survey off the west coast of North America, 40°N latitude to 52°N latitude. *Bull. Geol. Soc. Am.,* 72: 1267–1270.

Raleigh, C. B., 1967a. Tectonic implications of serpentine weakening. *Geophys. J. Roy. Astron. Soc. London,* 13: 113–118.

Raleigh, C. B., 1967b. Experimental deformation of ultramafic rocks and minerals. In: P. J. Wyllie (Editor), *Ultramafic and Related Rocks.* Wiley, New York, N.Y., pp.191–199.

Raleigh, C. B. and Kirby, S. H., 1970. Creep in the upper mantle. *Mineral. Soc. Am., Spec. Papers,* 3: 113–121.

Raleigh, C. B. and Paterson, H. S., 1965. Experimental deformation of serpentirite and its tectonic implication. *J. Geophys. Res.,* 70: 3965–3985.

Rayleigh, Lord, 1916. On convective currents in a horizontal layer of fluid when the high temperature is on the underside. *Phil. Mag.,* 32: 529–546.

Research Group for Explosion Seismology, 1951–1966. Explosion-seismic observations in north-eastern Japan, 1–5; Crustal structure in northern Kwanto district by explosion-seismic observations; Observations of seismic waves for explosions, 1–2; Crustal structure derived from explosion-seismic observations, 1–3; crustal structure in the western part of Japan derived from the observation of the first and second Kurayosi and the Hanabusa explosions. *Bull. Earthquake Res. Inst.,* 29: 97–106; 30: 279–292; 31: 281–289; 32: 79–86; 33: 699–707; 36: 329–348; 37: 89–112; 37: 495–508; 39: 285–326; 42: 515–531; 44: 89–107.

Research Group for Late Mesozoic Igneous Activity of Southwest Japan, 1967. Late Mesozoic igneous activity and tectonic history in the inner zone of southwest Japan. *Assoc. Geol. Collabor. Japan, Monogr.,* 13: 1–50. (In Japanese with English abstract.)

Research Group for Quaternary Tectonic Maps, 1968. Quaternary tectonic map of Japan. *Quaternary Res.,* 7: 182–187. (In Japanese with English abstract.)

Research Group for the Travel Time Curve, 1968. Travel time curve of near earthquakes in the Japan area, and some related problems. Procedures and preliminary results. *Bull. Earthquake Res. Inst.,* 45: 625–656.

Richter, C. F., 1960. Comparison of block and arc tectonics in Japan with those of some other regions. *J. Phys. Earth,* 8: 1–10.

Rikitake, T., 1952. Electrical conductivity and temperatures in the earth. *Bull. Earthquake Res. Inst.,* 30: 13–24.

Rikitake, T., 1956. Electrical state and seismicity beneath Japan. *Bull. Earthquake Res. Inst.,* 34: 291–300.

Rikitake, T., 1959a. Anomaly of geomagnetic variations in Japan. *Geophys. J. Roy. Astron. Soc. London,* 2: 276–287.

Rikitake, T., 1959b. Geophysical evidence of the olivine–spinel transition hypothesis in the earth's mantle. *Bull. Earthquake Res. Inst.,* 37: 423–431.

Rikitake, T., 1959c. The anomalous behaviour of geomagnetic variations of a short period in Japan and its relation to the subterranean structure. *Bull. Earthquake Res. Inst.,* 37: 545–570.

Rikitake, T., 1966. *Electromagnetism and the Earth's Interior.* Elsevier, Amsterdam, 308 pp.

Rikitake, T., 1969. The undulation of an electrically conductive layer beneath the islands of Japan. *Tectonophysics,* 7: 257–264.

Rikitake, T. and Horai, K., 1960. Terrestrial heat flows related to possible geophysical event. *Bull. Earthquake Res. Inst.,* 38: 403–419.

Rikitake, T., Yokoyama, I. and Hishiyama, Y., 1952. A preliminary study on the anomalous behaviour of geomagnetic variations of short period in Japan and its relation to the subterranean structure. *Bull. Earthquake Res. Inst.,* 30: 207–221.

Rikitake, T., Yokoyama, I. and Hishiyama, Y., 1953. The anomalous behaviour of geomagnetic variations of short period in Japan and its relation to the subterranean structure. *Bull. Earthquake Res. Inst.,* 31: 19–31; 89–100; 101–118; 119–127.

Rikitake, T., Yokoyama, I. and Sato, S., 1956. Anomaly of the geomagnetic Sq variation in Japan and its relation to the subterranean structure. *Bull. Earthquake Res. Inst.,* 34: 197–235.

Rikitake, T., Yabu, T. and Yamakawa, K., 1962. The anomalous behaviour of geomagnetic variations of short period in Japan and its relation to the subterranean structure. *Bull. Earthquake Res. Inst.,* 40: 693–717.

Rikitake, T., Yokoyama, I., Uyeda, S., Yukutake, T. and Nakagawa, E., 1958. The anomalous behaviour of geomagnetic variations of short period in Japan and its relation to the subterranean structure. *Bull. Earthquake Res. Inst.,* 36: 1–20.

Rikitake, T., Uyeda, S., Yukutake, T., Tanaoka, I. and Nakagawa, E., 1959. The anomalous behaviour of geomagnetic variations of short period in Japan and its relation to the subterranean structure. *Bull. Earthquake Res. Inst.,* 37: 1–11.

Ringwood, A. E., 1966. Mineralogy of the mantle. In: P. M. Hurley (Editor), *Advances in Earth Science.* M.I.T. Press, Cambridge, Mass., pp.357–399.

Ritsema, A. R., 1961. Further forcal mechanism studies at De Bilt. *Publ. Dom. Obs. Ottawa,* 24: 355–358.

Roden, R. B., 1963. Electromagnetic core–mantle coupling. *Geophys. J. Astron. Soc. London,* 7: 361–374.

Runcorn, S. K., 1962. Towards a theory of continental drift. *Nature,* 193: 313–314.

Saito, M. and Takeuchi, H., 1966. Surface waves across the Pacific. *Bull. Seismol. Soc. Am.,* 56: 1067–1091.

Sakamoto, T., 1966. Cenozoic strata and structural development in the southern half of the Toyama Basin, central Japan. *Chishitsu Chosasho Hokoku,* 213: 1–27. (In Japanese with English abstract.)

Santo, T., 1961. Division of the southwestern Pacific area into several regions in each of which Rayleigh waves have the same dispersion characteristics. *Bull. Earthquake Res. Inst.,* 39: 603–630.

Santo, T., 1963. Division of the Pacific area into seven regions in each of which Rayleigh waves have the same group velocities. *Bull. Earthquake Res. Inst.,* 41: 719–741.

Santo, T., 1969a. Characteristics of seismicity in South America. *Bull. Earthquake Res. Inst.,* 47: 635–672.

Santo, T., 1969b. Regional study on the characteristic seismicity of the world. 1. Hindu Kush region. *Bull. Earthquake Res. Inst.*, 47: 1035–1048.

Santo, T., 1969c. Regional study on the characteristic seismicity of the world. 2. From Burma down to Java. *Bull. Earthquake Res. Inst.*, 47: 1049–1061.

Santo, T., 1970. Regional study on the characteristic seismicity of the world, 3. New Hebrides Islands region. *Bull. Earthquake Res. Inst.*, 48: 1–18.

Santo, T., and Sato, Y., 1966. World-wide survey of the regional characteristics of group velocity dispersion of Rayleigh waves. *Bull. Earthquake Res. Inst.*, 44: 939–964.

Sasai, Y., 1968. Effect of sea on geomagnetic changes. *Proc. Symp. Conductivity Anomaly – Earthquake Res. Inst., Univ. Tokyo*, pp.133–142. (In Japanese.)

Sasajima, S., Nishida, J. and Shimada, M., 1968. Paleomagnetic evidence of a drift of the Japanese main island during the Paleogene Period. *Earth Planetary Sci. Lett.*, 5: 135–221.

Sassa, K., 1951. *Great Earthquakes*. Kobundo, Tokyo. (In Japanese.)

Sato, T., 1969. *Submarine Topography*. Lattice, Tokyo, 191 pp. (In Japanese.)

Sato, T. and Ono, K., 1964. The submarine geology off San'in district, southern Japan Sea. *J. Geol. Soc. Japan*, 70: 434–445. (In Japanese with English abstract.)

Savage, J. C. and Hastie, L. M., 1966. Surface deformation associated with dip-slip faulting. *J. Geophys. Res.*, 71: 4897–4904.

Schmucker, U., 1964. Anomalies of geomagnetic variations in the southwestern United States. *J. Geomagn. Geoelectr.*, 15: 193–221.

Schmucker, U., Hartmann, O., Giesecke Jr., A. A., Casaverde, M. and Forbush, S. E., 1964. Electrical conductivity anomalies in the earth's crust in Peru. *Carnegie Inst., Wash., Year Book*, 63: 354–362.

Sclater, J. G. and Francheteau, J., 1970. The implications of terrestrial heat flow observations on current tectonic and geochemical models of the crust and upper mantle of the earth. *Geophys. J.*, 20: 493–509.

Sclater, J. G. and Menard, H. W., 1967. Topography and heat flow of the Fiji Plateau. *Nature*, 216: 991–993.

Segawa, J., 1968. Measurement of gravity at sea around Japan, 1967. *J. Geod. Soc. Japan*, 13: 53–65.

Segawa, J., Ozawa, K. and Tomoda, Y., 1967. Local magnetic anomalies in the north Pacific Ocean. *La Mer*, 5: 8–20.

Seki, Y., Oki, Y., Matsuda, T., Mikami, K. and Okumura, K., 1969. Metamorphism in the Tanzawa Mountains, central Japan. *J. Japan Assoc. Mineral. Petrol. Econ. Geologists*, 61: 1–75.

Seyfert, C. K., 1969. Undeformed sediments in oceanic trenches with sea-floor spreading. *Nature*, 222: 70.

Shimada, M., 1966a. Melting of a basalt at high pressure. *Zisin*, 19: 167–175. (In Japanese with English abstract.)

Shimada, M., 1966b. Effects of pressure and water on the melting of basalts. *Zisin*, 19. (In Japanese with English abstract.)

Shimazu, M. and Sato, A., 1966. On the diagenetic alteration of the volcanic rocks in the Ainai–Jimba area, northern Akita district, northeastern Japan. *J. Fac. Sci., Niigata Univ.*, 1: 59–95. (In Japanese with English abstract.)

Shimazu, Y., 1961. Physical theory of generation, upward transfer, differentiation, solidification and explosion of magmas. *J. Earth Sci., Nagoya Univ.*, 9: 185–223.

Shimazu, Y. and Kohno, Y., 1964. Unsteady mantle convection and tectogenesis. *J. Earth Sci., Nagoya Univ.*, 12: 102–115.

Shimozuru, D., 1963. On the possibility of the existence of the molten portion in the upper mantle of the earth. *J. Phys. Earth.*, 11: 49–55.

Shor Jr., G. G., 1966. Continental margins and island arcs of western North America. *Geol. Surv. Can., Paper*, 66: 216–222.

Simmons, G., 1966. Heat flow in the earth. *J. Geol. Educ.*, 14: 105–110.

Simmons, G. and Horai, K., 1968. Heat flow data, 2. *J. Geophys. Res.*, 73: 6608–6629.

Sollas, W. J., 1903. The figure of the earth. *Quart. J. Geol. Soc. London,* 59: 180–188.

Stauder, W., 1968a. Tensional character of earthquake foci beneath the Aleutian Trench with relation to sea-floor spreading. *J. Geophys. Res.,* 73: 7693–7701.

Stauder, W., 1968b. Mechanism of the Rat Island earthquake sequence of February 4, 1965, with relation to island arcs and sea-floor spreading. *J. Geophys. Res.,* 73: 3847–3858.

Stauder, W. and Bollinger, G. A., 1964. The S-wave project for focal mechanism studies: earthquakes of 1962. *Bull. Seismol. Soc. Am.,* 54: 2198–2208.

Stauder, W. and Bollinger, G. A., 1966a. The S-wave project for focal mechanism studies: earthquakes of 1963. *Bull. Seismol. Soc. Am.,* 56: 1363–1372.

Stauder, W. and Bollinger, G. A., 1966b. The focal mechanism of the Alaska earthquake of March 28, 1964, and of its aftershock sequence. *J. Geophys. Res.,* 71: 5283–5296.

Stauder, W. and Udias, A. S. J., 1963. S-wave studies of earthquakes of the north Pacific, 2. Aleutian Islands. *Bull. Seismol. Soc. Am.,* 53: 59–77.

Stille, H., 1955. Recent deformations of the earth's crust in the light of those of earlier epochs. *Geol. Soc. Am., Spec. Papers,* 62: 171–192.

Suggate, R. P., 1965. The tempo of events in New Zealand geological history. *N.Z. J. Geol. Geophys.,* 8: 1139–1148.

Sugimura, A., 1952. On the deformation of the earth's surface owing to the rock folding in Japan. *Bull. Earthquake Res. Inst.,* 30: 163–178. (In Japanese with English abstract.)

Sugimura, A., 1958. Marianen–Japan–Kamtschatka Orogen in Tätigkeit. *Earth Sci.,* 37: 34–39. (In Japanese with German abstract.)

Sugimura, A., 1960. Zonal arrangement of some geophysical and petrological features in Japan and its environs. *J. Fac. Sci., Univ. Tokyo, Sect. 2,* 12: 133–153.

Sugimura, A., 1961. Regional variation of the K_2O/Na_2O ratios of volcanic rocks in Japan and environs. *J. Geol. Soc. Japan,* 67: 292–300. (In Japanese with English abstract.)

Sugimura, A., 1962a. Two examples of Recent and Pleistocene crustal movements in Japan. In: O. Meisser (Editor), *Intern. Symp. on Recent Crustal Movement, 1st, Leipzig.* Akademie-Verlag, Berlin, pp.252–256.

Sugimura, A., 1962b. Japanese volcanic belts. *Geography,* 7: 1315–1321. (In Japanese.)

Sugimura, A., 1964. A relation of distribution of tilted terraces and folded terraces with geological structure. *Proc. U.S. Japan Conf. Res. Relat. Earthquake Prediction Probl.* Earthquake Res. Inst., Tokyo, pp.98–99.

Sugimura, A., 1966. Complementary distributions of epicenters of mantle earthquakes and of loci of volcanoes in island arcs. *Zisin,* 19: 96–106. (In Japanese with English abstract.)

Sugimura, A., 1967a. Uniform rates and duration period of Quaternary earth movements in Japan. *J. Geosci., Osaka City Univ.,* 10(1/4): 25–35.

Sugimura, A., 1967b. Chemistry of volcanic rocks and seismicity of the earth's mantle in the island arcs. *Bull. Volcanol.,* 30: 319–334.

Sugimura, A., 1968. Spatial relations of basaltic magmas in island arcs. In: H. H. Hess and A. Poldervaart (Editors), *Basalts: the Poldervaart Treatise on Rocks of Basaltic Composition.* Interscience, New York, N.Y., 2: 537–571.

Sugimura, A., 1971. Quaternary tectonics in Japan. In: M. Ters (Editor), *Etudes sur le Quaternaire dans le Monde – Inqua Congr., 8th, Paris,* pp. 847–850.

Sugimura, A. and Matsuda, T., 1965. Atera fault and its displacement vectors. *Geol. Soc. Am., Bull.,* 76: 509–522.

Sugimura, A. and Naruse, Y., 1954. Changes in sea level, seismic upheavals, and coastal terraces in the southern Kanto region, Japan, 1. *Japan J. Geol. Geogr.,* 24: 101–113.

Sugimura, A. and Uyeda, S., 1967. A possible anisotropy of the upper mantle accounting for deep earthquake faulting. *Tectonophysics,* 5: 25–33.

Sugimura, A., Matsuda, T., Chinzei, T. and Nakamura, K., 1963. Quantitative distribution of Late Cenozoic volcanic materials in Japan. *Bull. Volcanol.,* 26: 125–140.

Suzuki, T., 1966. On the distribution of volcanoes in relation to the configurations of their basal rocks in the circum-Pacific orogenic zone. *Proc. Pacific Sci. Congr., 11th, Tokyo: Physiographic*

Development of the Pacific Region, p.13.

Sykes, L. R., 1966. The seismicity and deep structure of island arcs. *J. Geophys. Res.,* 71: 2981–3006.

Sykes, L. R., 1967. Mechanism of earthquakes and nature of faulting on the mid-oceanic ridges. *J. Geophys. Res.,* 72: 2131–2153.

Sykes, L. R. and Ewing, M., 1965. The seismicity of the Carribbean region. *J. Geophys. Res.,* 70: 5065–5074.

Sykes, L. R., Isacks, B. L. and Oliver, J., 1969. Spatial distribution of deep and shallow earthquakes of small magnitudes in the Fiji–Tonga region. *Bull. Seismol. Soc. Am.,* 51: 1093–1113.

Takai, F., Matsumoto, T. and Toriyama, R. (Editors), 1963. *Geology of Japan.* Univ. Tokyo Press, Tokyo, 279 pp.

Takeuchi, H. and Hasegawa, Y., 1967. Viscosity distribution within the earth. *Geophys. J. Roy. Astron. Soc. London,* 9: 503–508.

Takeuchi, H. and Kanamori, H., 1968. Crustal deformations before and after great earthquakes and mantle convection. *J. Seismol. Soc. Japan,* 21: 316. (In Japanese.)

Takeuchi, H. and Sakata, M., 1970. Convection in a mantle with variable viscosity. *J. Geophys. Res.,* 75: 921–927.

Takeuchi, H. and Uyeda, S., 1965. A possibility of present day regional metamorphism. *Tectonophysics,* 2: 59–68.

Takeuchi, H., Hamano, Y. and Hasegawa, Y., 1968. Raleigh- and Love-wave discrepancy and the existence of magma pockets in the upper mantle. *J. Geophys. Res.,* 73(10): 3349–3350.

Talwani, M., Sutton, G. H. and Worzel, J. L., 1959. Crustal section across the Puerto Rico Trench. *J. Geophys. Res.,* 64: 1545–1555.

Talwani, M., Worzel, J. L. and Ewing, M., 1961. Gravity anomalies and crustal section across the Tonga Trench. *J. Geophys. Res.,* 66: 1265–1278.

Talwani, M., Le Pichon, X. and Ewing, M., 1965. Crustal structure of the mid-ocean ridges, 2. Computed model from gravity and seismic refraction data. *J. Geophys. Res.,* 70: 341–352.

Tayama, R., 1950. The submarine configuration off Sikoku, especially of the continental slope. *Hydrograph. Bull.,* Spec. Number, 7: 54–82. (In Japanese with English abstract.)

Tazima, M., 1966a. Measurement of gravity at sea. *Kokudo-chiriin Jiho,* 31: 10–15. (In Japanese.)

Tazima, M., 1966b. *Accuracy of Recent Magnetic Survey and a Locally Anomalous Behaviour of the Geomagnetic Secular Variation in Japan.* Thesis, Univ. of Tokyo, Tokyo.

Terada, T., 1934. On bathymetrical features of the Japan Sea. *Bull. Earthquake Res. Inst.,* 12: 650–656.

Thellier, E., 1951. Propriétés magnétiques des terres cuites et des roches. *J. Phys. Radium,* 12: 208–218.

Tilley, C. E., 1950. Some aspects of magmatic evolution. *Quart. J. Geol. Soc. London,* 106: 37–61.

Tokuda, S., 1926. On the echelon structure of the Japanese Archipelago. *Japan. J. Geol. Geogr.,* 5: 41–76.

Tokuoka, T., 1967. The Shimanto terrain in the Kii Peninsula, southwest Japan. *Mem. Fac. Sci., Kyoto Univ., Ser. Geol. Mineral.,* 34: 35–74.

Tomita, T., 1932. Geological and petrological study of Dogo, Oki, 19 and 20. *J. Geol. Soc. Tokyo,* 39: 609–640; and 675–691. (In Japanese.)

Tomita, T., 1935. On the chemical compositions of the Cenozoic alkaline suite of the circum-Japan Sea region. *J. Shanghai Sci. Inst., Sect. II,* 1: 227–306.

Tomoda, Y. and Segawa, J., 1967. Measurement of gravity and total magnetic force in the sea near and around Japan, 1966. *J. Geod. Soc. Japan,* 12: 157–164.

Tomoda, Y., Ozawa, K. and Segawa, J., 1968. Measurement of gravity and magnetic field on board of a cruising vessel. *Bull. Oceanogr. Res. Inst., Univ. Tokyo,* 3: 1.

Tomoda, Y., Segawa, J. and Tokuhiro, A., 1970. Free-air gravity anomaly around Japan. *Proc. Japan Acad.,* 46: 1006–1010.

Tozer, D., 1959. The electrical properties of the earth's interior. In: L. H. Ahrens, F. Press, K. Rankama and S. K. Runcorn (Editors), *Physics and Chemistry of the Earth*. Pergamon, London, 3: 414–436.

Tritton, D. J. and Zarraga, M. N., 1967. Convection in a horizontal layer with internal heat generation. *J. Fluid Mech.*, 30: 21–31.

Tsuboi, C., 1954. Gravity survey along the lines of precise levels throughout Japan by means of a Worden gravimeter, 4. Map of Bouguer anomaly distribution in Japan based on approximately 4,500 measurements. *Bull. Earthquake Res. Inst., Suppl.*, 4: 125–127.

Tsuboi, C., 1964. Time rate of energy release by earthquakes in and near Japan. *J. Phys. Earth*, 12: 25–36.

Tsuboi, C., 1965. Time rate of earthquake energy release in and around Japan. *Proc. Japan Acad.*, 41: 392–397.

Tsuboi, C., Jitsukawa, A., Tajima, H. and Okada, A., 1953. Gravity survey along the lines of precise levels throughout Japan by means of a Worden gravimeter, 1. Shikoku district. *Bull. Earthquake Res. Inst., Suppl.*, 4: 1–45.

Tsuboi, C., Jitsukawa, A., Tajima, H. and Okada, A., 1954a. Gravity survey along the lines of precise levels throughout Japan by means of a Worden gravimeter, 2. Chugoku district; and 3. Supplement to the previous report of the gravity survey in Shikoku. *Bull. Earthquake Res. Inst., Suppl.*, 4: 47–123.

Tsuboi, C., Jitsukawa, A., Tajima, H. and Okada, A., 1954b. Gravity survey along the lines of precise levels throughout Japan by means of a Worden gravimeter, 5. Kinki district. *Bull. Earthquake Res. Inst., Suppl.*, 4: 129–198.

Tsuboi, C., Jitsukawa, A., Tajima, H. and Okada, A., 1955. Gravity survey along the lines of precise levels throughout Japan by means of a Worden gravimeter, 6. Chubu district. *Bull. Earthquake Res. Inst., Suppl.*, 4: 199–310.

Tsuboi, C., Jitsukawa, A., Tajima, H. and Okada, A., 1956. Gravity survey along the lines of precise levels throughout Japan by means of a Worden gravimeter, 7. Tohoku district; 8. Kanto district; and 9. Kyushu district. *Bull. Earthquake Res. Inst., Suppl.*, 4: 311–406; 407–474; and 476–552.

Tsuboi, C., Tomoda, Y. and Kanamori, H., 1961. Continuous measurements or gravity on board a moving surface ship. *Proc. Japan. Acad.*, 37: 571–576.

Tsuchi, R., 1961. On the Late Neogene sediments and molluscs in the Tokai region, with notes on the geologic history of the Pacific coast of southwest Japan. *Japan. J. Geol. Geogr.*, 32: 437–456.

Tsuchi, R. and Kagami, H., 1967. Discovery of nerineid gastropods from seamount Sisoev (Erimo) at the junction of Japan and Kurile–Kamchatka trenches. *Rec. Oceanogr. Works, Japan*, 9: 1–6.

Tsumura, K., 1964. Japanese contribution to the study of mean sea-level (a review). *J. Geod. Soc. Japan*. 10: 192–202.

Tsuneishi, Y., 1966. Geologic structure of the Hirono area in the Abukuma Mountains. *Bull. Earthquake Res. Inst.*, 44: 749–764. (In Japanese with English abstract.)

Turcotte, P. L. and Oxburgh, E. R., 1967. Finite amplitude convective cells and continental drift. *J. Fluid Mech.*, 28: 29–42.

Turcotte, P. L. and Oxburgh, E. R., 1968. A fluid theory for the deep structure of dip-slip fault zones. *Phys. Earth Planetary Interior*, 1: 381–386.

Turcotte, P. L. and Oxburgh, E. R., 1969. Convection in a mantle with variable physical properties. *J. Geophys. Res.*, 74: 1458–1474.

Turner, F. J., 1942. Preferred orientation of olivine crystals in peridotites, with special reference to New Zealand samples. *Roy. Soc. New Zealand, Trans.*, 72(3): 280–300.

Tuyezov, I. K., Sichev, P. M., Tarakanov, R. Z. and Krasny, M. L., 1968. Structure of the folded areas and recent geosynclines of the Okhotsk area. In: L. Knopoff, C. Drake and P. Hart (Editors), *The Crust and the Upper Mantle of the Pacific Area – Am. Geophys. Union, Geophys. Monogr.*, 12: 473–480.

Udias, A. S. J. and Stauder, W. S. J., 1964. Application of numerical method for S-wave focal mechanism determination to earthquakes of the Kurile–Kamchatka Islands region. *Bull.*

Seismol. Soc. Am., 54: 2049—2066.

Uffen, R. J., 1952. A method of estimating the melting point gradient in the earth's mantle. *Trans. Am. Geophys. Union*, 33: 892—896.

Umbgrove, J. H. F., 1938. Geological history of the East Indies. *Bull. Am. Assoc. Petrol. Geologists*, 22: 1—70.

Umbgrove, J. H. F., 1947. *The Pulse of the Earth*. Nijhoff, The Hague.

Ustiyev, Y. K., 1963. Problems of volcanism and plutonism; volcano-plutonic formations. *Intern. Geol. Rev.*, 7: 1994—2016.

Utada, M., 1965. Zonal distribution of authigenic zeolites in the Tertiary pyroclastic rocks in Mogami district, Yamagata prefecture. *Sci. Papers, Coll. Gen. Educ., Univ. Tokyo*, 15: 173—216.

Utsu, T., 1966. Regional differences in absorption of seismic waves in the upper mantle as inferred from abnormal distributions of seismic intensities. *J. Fac. Sci., Hokkaido Univ., Ser. VII*, 2: 359—374.

Utsu, T., 1967. Anomalies in seismic wave velocity and attenuation associated with a deep earthquake zone, 1. *J. Fac. Sci., Hokkaido Univ., Ser. VII*, 3: 1—25.

Utsu, T. and Okada, H., 1968. Anomalies in seismic wave velocity and attenuation associated with a deep earthquake zone, 2. *J. Fac. Sci., Hokkaido Univ., Ser. VII.*, 3: 65—84.

Uyeda, S., 1958. Thermo-remanent magnetism as a medium of palaeomagnetism, with special reference to reverse thermo-remanent magnetism. *Japan J. Geophys.*, 2: 1—123.

Uyeda, S. and Horai, K., 1960. Terrestrial heat flow at Innai oil field, Akita prefecture and at three localities in Kanto district, Japan. *Bull. Earthquake Res. Inst.*, 38: 421—436.

Uyeda, S. and Horai, K., 1963a. Terrestrial heat flow measurements in Kanto and Chubu districts, Japan. *Bull. Earthquake Res. Inst.*, 41: 83—107.

Uyeda, S. and Horai, K., 1963b. Terrestrial heat flow measurements in Kinki, Chugoku and Shikoku districts, Japan. *Bull. Earthquake Res. Inst.*, 41: 109—135.

Uyeda, S. and Horai, K., 1964. Terrestrial heat flow in Japan. *J. Geophys. Res.*, 69: 2121—2141.

Uyeda, S. and Richards, M., 1966. Magnetization of four Pacific seamounts near the Japanese Islands. *Bull. Earthquake Res. Inst.*, 44: 179—213.

Uyeda, S. and Rikitake, T., 1970. Electrical conductivity anomaly and terrestrial heat flow. *J. Geomagn. Geoelectr.*, 22: 75—90.

Uyeda, S. and Vacquier, V., 1968. Geomagnetic and geothermal data in and around Japan. In: L. Knopoff, C. Drake and P. Hart (Editors), *The Crust and Upper Mantle of the Pacific Area — Am. Geophys. Union, Geophys. Monogr.*, 12: 349—366.

Uyeda, S. and Watanabe, T., 1970. Preliminary report of terrestrial heat flow study in the South American continent; distribution of geothermal gradients. *Tectonophysics*, 10: 235—242.

Uyeda, S., Yukutake, T. and Tanaoka, I., 1958. Preliminary report of terrestrial heat flow in Japan. *Bull. Earthquake Res. Inst.*, 36: 251—273.

Uyeda, S., Tomoda, Y., Horai, K., Kanamori, H. and Futi, H., 1961. A sea bottom thermogradmeter. *Bull. Earthquake Res. Inst.*, 39: 115—131.

Uyeda, S., Sato, T., Yasui, M., Yabu, T., Watanabe, T., Kawada, K. and Hagiwara, Y., 1964. Reporter on geomagnetic survey in the northwestern Pacific during JEDS-7 and JEDS-8 cruises. *Bull. Earthquake Res. Inst.*, 42: 555—570.

Uyeda, S., Yasui, M., Sato, T., Akamatsu, H. and Kawada, K., 1965. Heat flow measurements during the JEDS-6 and JEDS-7 cruise in 1963. *Oceanogr. Mag.*, 16: 7—10.

Uyeda, S., Vacquier, V., Yasui, M., Sclater, J., Sato, T., Lawson, J., Watanabe, T., Dixon, F., Silver, E., Fukao, Y., Sudo, K., Nishikawa, M. and Tanaka, T., 1967. Results of geomagnetic survey during the cruise of R.V. "Argo" in the western Pacific 1966 and the compilation of magnetic charts of the same area. *Bull. Earthquake Res. Inst.*, 45: 799—814.

Vacquier, V. and Taylor, P. T., 1966. Geothermal and magnetic survey off the coast of Sumatra, 1. Presentation of data. *Bull. Earthquake Res. Inst.*, 44: 531—540.

Vacquier, V. and Uyeda, S., 1967. Palaeomagnetism of nine seamounts in the western Pacific and of three volcanoes in Japan. *Bull. Earthquake Res. Inst.*, 45: 815—848.

Vacquier, V., Raff, A. D. and Warren, R. E., 1961. Horizontal displacements in the floor of the northeastern Pacific Ocean. *Bull. Geol. Soc. Am.*, 72: 1251–1258.

Vacquier, V., Sclater, J. C. and Corry, C., 1967b. Heat flow in the eastern Pacific Ocean. *Bull. Earthquake Res. Inst.*, 45: 375–393.

Vacquier, V., Uyeda, S., Yasui, M., Sclater, J., Corry, C. and Watanabe, T., 1967a. Heat flow measurements in the northwestern Pacific. *Bull. Earthquake Res. Inst.*, 44: 1519–1535.

Vening Meinesz, F. A., 1930. Maritime gravity surveys in the Netherlands East Indies; tentative interpretation of the results. *Koninkl. Ned. Akad. Wetenschap., Proc.*, 33: 566–577.

Vening Meinesz, F. A., 1962. Thermal convection in the earth's mantle. In: S. K. Runcorn (Editor), *Continental Drift*. Academic Press, New York, N.Y., pp.145–176.

Vening Meinesz, F. A., 1964. *The Earth's Crust and Mantle*. Elsevier, Amsterdam.

Vening Meinesz, F. A., Umbgrove, J. H. F. and Kuenen, P. H., 1934. *Gravity Expeditions at Sea, 2: 1923–1930*. Waltman, Delft, 208 pp.

Verhoogen, J., 1946. Volcanic heat. *Am. J. Sci.*, 244: 745–771.

Verhoogen, J., 1954. Petrological evidence on temperature distribution in the mantle of the earth. *Trans. Am. Geophys. Union*, 35: 85–92.

Verma, R. K., 1960. Elasticity of some high-density crystals. *J. Geophys. Res.*, 65: 757–766.

Vine, F. J., 1966. Spreading of the ocean floor. *Science*, 154: 1405–1415.

Vine, F. J. and Matthews, D. H., 1963. Magnetic anomalies over oceanic ridges. *Nature*, 199: 947–949.

Vine, F. J. and Wilson, J. T., 1965. Magnetic anomalies over a young oceanic ridge off Vancouver Island. *Science*, 150: 485–489.

Von Herzen, R. P. and Uyeda, S., 1963. Heat flow through the eastern Pacific Ocean floor. *J. Geophys. Res.*, 68: 4219–4250.

Wadati, K., 1935. On the activity of deep-focus earthquakes in the Japan Island and neighbourhoods. *Geophys. Mag.*, 8: 305–325.

Wadati, K., Sagisaka, K. and Masuda, K., 1933. On the travel time of earthquake waves, 1. *Geophys. Mag.*, 7: 87–99.

Watanabe, T., 1968. Temperature profiles at continental margins. *Proc. Symp. on Conductivity Anomaly, Earthquake Res. Inst., Univ. Tokyo*, pp.167–182. (In Japanese.)

Watanabe, T., Epp, D., Uyeda, S., Langseth, M. and Yasui, M., 1970. Heat flow in the Philippine Sea. *Tectonophysics*, 10: 205–224.

Weertman, J., 1970. The creep strength of the earth's mantle. *Rev. Geophys. Space Phys.*, 8: 145–168.

Wilson, J. T., 1959. Geophysics and continental growth. *Am. Scientist*, 47: 1–24.

Wilson, J. T., 1963. Continental drift. *Sci. Am.*, April, 1963: 1–16.

Wilson, J. T., 1965. A new class of faults and their bearing on continental drift. *Nature*, 207: 343–347.

Woollard, G. P. and Strange, W. E., 1962. Gravity anomalies and the crust of the earth in the Pacific Basin. In: G. A. MacDonald and H. Kuno (Editors), *The Crust of the Pacific Basin – Am. Geophys. Union, Geophys. Monogr.*, 6: 60–80.

Worzel, J. L., 1965a. *Pendulum Gravity Measurements at Sea, 1936–1959*. Wiley, New York, N.Y.

Worzel, J. L., 1965b. Structure of continental margins and development of ocean trenches. Continental margins and island arcs. *Geol. Surv. Can., Paper*, 66–15: 357–375.

Worzel, J. L. and Ewing, M., 1952. Gravity measurements at sea, 1948 and 1949. *Trans. Am. Geophys. Union*, 33: 453–460.

Worzel, J. L. and Shurbet, G. I., 1955. Gravity interpretations from standard and continental sections. *Geol. Soc. Am., Spec. Papers*, 62: 87–100.

Yabe, H., 1918. Itoigawa–Shizuoka fault. *Gendai-no-Kagaku (Mod. Sci.)*, 6: 1–4. (In Japanese.)

Yamada, T., Kawachi, Y., Watanabe, T., Yokota, Y. and Kwanke, E., 1969. The Shimanto belt of the northern Akaishi Mountainlands, central Japan. *Mem. Geol. Soc. Japan,* 4: 117–122. (In Japanese with English abstract.)

Yamagata, O., Yonechi, F., Suzuki, M. and Sugimura, A., 1973. Radiocarbon dates from Mogami district in northeast Japan. *Earth Sci.* (In Japanese, in preparation.)

Yasui, M. and Watanabe, T., 1965. Terrestrial heat flow in the Japan Sea, 1. *Bull. Earthquake Res. Inst.,* 43: 549–563.

Yasui, M., Kishii, T. and Sudo, K., 1967a. Terrestrial heat flow in the Okhotsk Sea, 1. *Oceanogr. Mag.,* 19: 87–94.

Yasui, M., Hashimoto, Y. and Uyeda, S., 1967c. Geomagnetic studies of the Japan Sea, 1. Anomaly pattern in the Japan Sea. *Oceanogr. Mag.,* 19: 221–231.

Yasui, M., Hashimoto, Y. and Uyeda, S., 1967d. Geomagnetic and bathymetric study of the Okhotsk Sea, 1. *Oceanogr. Mag.,* 19: 73–85.

Yasui, M., Isezaki, N. and Uyeda, S., 1972. Paleomagnetism of seamounts around Japan. (In preparation.)

Yasui, M., Horai, K., Uyeda, S. and Akamatsu, H., 1963. Heat flow measurements in the western Pacific during the JEDS-5 and other cruises in 1962 aboard M.S. "Ryofu Maru". *Oceanogr. Mag.,* 14(2): 147–156.

Yasui, M., Watanabe, T., Uyeda, S. and Kishii, T., 1967b. Terrestrial heat flow in the Japan Sea, 2. *Bull. Earthquake Res. Inst.,* 44: 1501–1518.

Yasui, M., Kishii, T., Watanabe, T. and Uyeda, S., 1968a. Heat flow in the Japan Sea. In: L. Knopoff, C. Drake and P. Hart (Editors), *The Crust and Upper Mantle of the Pacific Area — Am. Geophys. Union,* 12: 3–16.

Yasui, M., Nagasaka, K., Kishii, T. and Halunen, A. J., 1968b. Terrestrial heat flow in the Okhotsk Sea, 2. *Oceanogr. Mag.,* 20: 73–86.

Yasui, M., Epp, D., Nagasaka, K. and Kishii, T., 1970. Terrestrial heat flow in the seas round the Nansei Shoto (Ryukyu Islands). *Tectonophysics,* 10: 225–234.

Yoder Jr., H. S., 1969. Calc-alkalic andesites: experimental data bearing on the origin of their assumed characteristics. In: A. R. McBirney (Editor), *Proceedings of the Andesite Conference (Eugene and Bond).*

Yoder Jr., H. S. and Tilley, C. E., 1962. Origin of basalt magmas; an experimental study of natural and synthetic rock systems. *J. Petrol.,* 3: 342–532.

Yokoyama, I., 1956–1957. Energetics in active volcanoes, 1–3. *Bull. Earthquake Res. Inst.,* 34: 190–195; 35: 75–98; and 35: 99–108.

Yoshikawa, T., Kaizuka, S. and Ota, Y., 1964. Crustal movement in the Late Quaternary revealed with coastal terraces on the southeast coast of Shikoku, southwestern Japan. *J. Geodet. Soc. Japan,* 10(3–4): 116–122.

Yoshikawa, T., Kaizuka, S. and Ota, Y., 1968. Coastal development of the Japanese Islands. *Proc. INQUA Congr., 7th, Utah.* Univ. Utah Press, Utah, 8: 457–465.

Yoshimura, T., 1961. Zeolites in the Miocene pyroclastic rocks in the Oshima–Fukushima district, southwestern Hokkaido. *J. Geol. Soc. Japan,* 67: 578–583. (In Japanese with English abstract.)

INDEX TO AUTHORS CITED IN TEXT

SUBJECT INDEX